Paying for Biodiversity

ENHANCING THE COST-EFFECTIVENESS
OF PAYMENTS FOR ECOSYSTEM SERVICES

OECD

ORGANISATION FOR ECONOMIC CO-OPERATION AND DEVELOPMENT

The OECD is a unique forum where governments work together to address the economic, social and environmental challenges of globalisation. The OECD is also at the forefront of efforts to understand and to help governments respond to new developments and concerns, such as corporate governance, the information economy and the challenges of an ageing population. The Organisation provides a setting where governments can compare policy experiences, seek answers to common problems, identify good practice and work to co-ordinate domestic and international policies.

The OECD member countries are: Australia, Austria, Belgium, Canada, Chile, the Czech Republic, Denmark, Finland, France, Germany, Greece, Hungary, Iceland, Ireland, Italy, Japan, Korea, Luxembourg, Mexico, the Netherlands, New Zealand, Norway, Poland, Portugal, the Slovak Republic, Slovenia, Spain, Sweden, Switzerland, Turkey, the United Kingdom and the United States. The European Commission takes part in the work of the OECD.

OECD Publishing disseminates widely the results of the Organisation's statistics gathering and research on economic, social and environmental issues, as well as the conventions, guidelines and standards agreed by its members.

ISBN 978-92-64-09026-2 (print)
ISBN 978-92-64-09027-9 (PDF)

Also available in French: *Payer pour la biodiversité : Améliorer l'efficacité-coût des paiements pour services écosystémiques*

Corrigenda to OECD publications may be found on line at: *www.oecd.org/publishing/corrigenda.*

Foreword

Biodiversity and associated ecosystem service loss and degradation present one of the major environmental challenges facing humankind. Despite the significant economic, social and cultural values they provide, such as food provisioning, clean water, genetic resources, climate regulation, and recreation benefits, biodiversity continues to be lost and in some areas at an accelerating rate. Given these trends, there is an urgent need for both (i) greater application of policies and incentives to address biodiversity conservation and sustainable use, and (ii) more efficient use of available finance in existing programmes. The latter is especially important in the context of the current economic crisis where public and private budgets are increasingly constrained and are competing with multiple demands.

The OECD Working Group on Economic Aspects of Biodiversity (WGEAB) has, for more than a decade, supported governments and institutions by providing analytical support on the valuation of biodiversity and ecosystem services, and the use of economic instruments, incentive measures and the creation of markets for the sustainable use and conservation of biological diversity.

This book, produced under the auspices of the WGEAB, considers an innovative mechanism known as Payments for Ecosystem Services (PES). PES are flexible, incentive-based mechanisms that have the potential to provide a cost-effective means of promoting the conservation and sustainable use of biodiversity and ecosystem services in a broad range of environmental, economic, and social contexts. Drawing on the literature and on practical experience from PES programmes in developed and developing countries, the book identifies good practice in the design and implementation of these programmes, with an emphasis on how to enhance their environmental and cost effectiveness.

This work is also of direct relevance to the Parties of the United Nations Convention on Biological Diversity (CBD), who requested further work in this area.

Acknowledgements

This book is a product of the Environment Directorate of the OECD. Preliminary versions of the chapters were presented at the meetings of the Working Group on Economic Aspects of Biodiversity (WGEAB) of the Environmental Policy Committee. Participants to these meetings provided valuable comments and suggestions. The book also builds on discussion at an OECD expert workshop on "Enhancing the Cost-effectiveness of Payments for Ecosystem Services" held on March 25, 2010. www.oecd.org/env/biodiversity

The book has been prepared by Katia Karousakis and Quiller Brooke, and has benefitted from helpful comments from OECD colleagues, namely Andrea Cattaneo, Jan Corfee-Morlot and Helen Mountford. Research assistance by Marie-Helene Greffard was provided on Chapter 3. Chapter contributions from Jim Binney and Charlie Zammit (Chapter 7) and Beria Leimonia and Brooke Kelsey Jack (Chapter 8) are gratefully acknowledged.

Jane Kynaston provided excellent administrative assistance and formatted the publication.

Table of contents

Acronyms and abbreviations 9

Executive summary 13

Introduction 21

Part I: Designing and implementing effective payments for ecosystem services programmes 35

Chapter 1: The economics of payments for ecosystem services 37

1.1 PES: an incentive-based mechanism 38
1.2 Spatial variability in the costs and benefits of biodiversity and ecosystem service provision 40
Notes 43
References 43

Chapter 2: Environmentally effective payments for ecosystem services 45

2.1 PES pre-requisites 46
2.2 General design elements for environmentally effective PES 48
Notes 54
References 54

Chapter 3: Cost-effective targeting of payments for ecosystem services 57

3.1 Targeting ecosystem services with high benefits 58
3.2 Targeting ecosystems services at risk of loss or degradation 64
3.3 Targeting providers with low opportunity costs 65
Notes 68
References 68

Chapter 4: Mobilising finance for payments for ecosystem services 71

4.1 Identifying ecosystem service financing needs and sources 72
4.2 Experience with private sector PES financing 79
Notes 87
References 88

Chapter 5: Insights for international payments for ecosystem services 93

5.1 Harnessing synergies between global carbon finance and biodiversity 94
5.2 International payments for biodiversity 95
Notes 97
References 97

Part II: Payments for ecosystem services programmes case studies 99

Chapter 6: United States: The USDA Conservation Reserve Programme 101

6.1 Introduction 102
6.2 The CRP general sign-up 106
6.3 The CRP environmental and cost effectiveness 112
6.4 Conclusions 124
Notes 124
References 125

Chapter 7: Australia: The Tasmanian Forest Conservation Fund 131

7.1 Introduction 132
7.2 Key design elements of the FCF 133
7.3 Effectiveness and efficiency of the FCF 141
7.4 Application of lessons in the Environmental Stewardship Programme 155
7.5 Conclusions 158
Notes 159
References 160

Chapter 8: Indonesia: A pilot PES auction in the Sumberjaya watershed 161

8.1 Introduction 162
8.2 Designing the PES inverse auction 163
8.3 Auction outcomes and environmental impacts 168
8.4 Conclusions 174
Notes 175
References 176

Chapter 9: Conclusions 179

References 187

Annex A: Case study overview 189

Tables

0.1. Policy approaches and instruments for biodiversity conservation and sustainable use 28
0.2. Annual PES budgets in selected national and regional PES programmes 29
4.1. Finance sources across a selection of PES programmes 80
6.1. Key elements of the CRP general sign-up auction 108
7.1. Key elements of the FCF 134
7.2. Area secured by the FCF 141
7.3. Duration of protection secured by the FCF 142
7.4. FCF inverse auction - key statistics 145
7.5. FCF inverse auction vs. other approaches 147
7.6. FCF management and administration costs 150
7.7. Potential conservation gains from the FCF auction approaches used 152
7.8. Analysis of CVI return on investment in the FCF 153
8.1. Indonesian pilot auction design characteristics 166
8.2. Indonesian pilot auction summary statistics 168
8.3. Rate of contract accomplishment in Indonesian pilot 172
9.1. Key design elements in three PES case studies 183
A.1. Case study overview 190

Figures

0.1. Four components of ecosystem services 23
0.2. The Total Economic Value 24
0.3. Costs of inaction with respect to biodiversity and ecosystem service loss 25
1.1. The PES mechanism: ecosystem beneficiaries pay the landholder supplying the service to compensate for the additional costs of conservation 38
1.2. Optimal provision of biodiversity, and the distributional and cost-effectiveness implications given a budget constraint 39
3.1. Targeting PES in Madagascar 60
3.2. Example of a UNEP-WCMC national map: Panama 61
3.3. Marketing biodiversity joint service provision 62
3.4. Targeting PEHS in Mexico 64
4.1. Stylistic representation of the spatial scale of different ecosystem service benefits 79
6.1. Acres of general CRP sign-up, 2009 104
6.2. The relative share of points awarded by category within all accepted bids, CRP general sign-up 1997 to 2003 117
6.3. Trends in USDA agri-environmental expenditures 123

7.1. Simplified representation of FCF implementation process 135
7.2. FCF supply curves for conservation 146
7.3. Conservation gains and relative costs between FCF inverse auctions vs. other approaches 149
8.1. Flow of the research steps in Indonesian pilot auction......................... 164
8.2. Supply curve resulting from Indonesian pilot auction.......................... 169
8.3. Average village compliance in Indonesian pilot within each site 171

Boxes

1.1. Design considerations in inverse auctions... 42
2.1. The integrated project management system for the Costa Rican PES..... 49
3.1. Metrics and indicators used to target biodiversity benefits in the Victorian BushTender and a Canadian pilot PES.................................. 59
3.2. Costa Rica Payments for Environmental Services 66
4.1. Forest-backed bonds for PES as part of sustainable forest management 74
4.2. Creating funds to finance PES in Ecuador and Tanzania....................... 78
6.1. The USDA portfolio of conservation programmes 104
6.2. The CRP continuous sign-up... 107
6.3. The CRP Environmental Benefits Index 110
6.4. Ecosystem services derived from wetland conservation in the Prairie Pothole Region 113

Acronyms and abbreviations

ARS	Agricultural Research Service
ARIES	Artificial Intelligence for Ecosystem Services
AUD	Australian dollar
BAU	Business as Usual
BBI	Biodiversity Benefits Index
CBD	Convention on Biological Diversity
CEAP	Conservation Effects Assessment Program
CITES	Convention on International Trade in Endangered Species
COP-9	9th meeting of the Conference of Parties
COP-10	10th meeting of the Conference of Parties
CRP	Conservation Reserve Program (United States)
CStP	Conservation Stewardship Program
CVI	Conservation Value Index
DAC	Development Assistance Committee
EAMCEF	Eastern Arc Mountains Conservation Endowment Fund (Tanzania)
EBI	Environmental Benefits Index
ESP	Environmental Stewardship Program
EQIP	Environmental Quality Incentives Program
EU	European Union
DSE	Department of Sustainability and Environment, Victoria, Australia
FACE	Forests Absorbing Carbon-dioxide Emissions Consortium
FCF	Forest Conservation Fund (Tasmania)

FSA	Farm Service Authority
FONAFIFO	*Fondo Nacional de Financiamiento Forestal* (National Forest Finance Fund – Costa Rica)
GEF	Global Environment Facility
GHG	Greenhouse Gases
GIS	Geographic Information System
HH	Habitat Hectare
HSS	Habitat Services Score
IAF	Inter-American Foundation
ICRAF	World Agroforestry Centre
IDR	Indonesian Rupiah
IIED	International Institute for Environment and Development
InVEST	Integrated Valuation of Ecosystem Services and Tradeoffs
IPES	International Payments for Ecosystem Services
IPMS	Integrated Project Management System
IUCN	International Union for Conservation of Nature
LP3ES	Institute of Social & Economic Research, Education & Information
NGO	Non-Governmental Organization
NRCS	Natural Resource Conservation Service
ODA	Official Development Assistance
OECD	Organisation for Economic Co-operation and Development
PES	Payments for Ecosystem Services
PEHS	Payments for Environmental Hydrological Services *(Pago de Services Ambientales Hydrologicas* - Mexico*)*
PSA	*Pago pro Servicios Ambientales* (Costa Rican PES)
REDD-plus	Reducing Emissions from Deforestation and Forest Degradation and conservation
RUPES	Rewarding the Upland Poor in Asia for Environmental Services
SLCP	Slopping Land Conversion Programme (China)
TEV	Total Economic Value

SOC	Soil Organic Carbon
USD	United States dollar
UNEP	United Nations Environmental Program
UNFAO	United Nations Food and Agricultural Organisation
UNFCCC	United Nations Framework Convention on Climate Change
USDA	United States Department of Agriculture
US EPA	United States Environmental Protection Agency
US GAO	United States General Accounting Office
US GS	United States Geological Survey
WCMC	World Conservation Monitoring Centre
WRP	Wetlands Reserve Program
WTA	Willingness to Accept
WWF	World Wildlife Fund

Executive summary

Biodiversity and ecosystems provide invaluable services to society. These include food, clean water, genetic resources, recreational services, flood protection, nutrient cycling and climate regulation, amongst many others. Ecosystem services provide critical life support functions and benefits, contributing to human health, security, well-being and economic growth. Despite the significant economic, social and cultural values of biodiversity and associated ecosystem services, biodiversity worldwide is being lost, and in some areas at an accelerating rate. Without renewed efforts to address this environmental challenge, OECD projections to 2030 indicate continued biodiversity loss.

Given these trends in biodiversity loss, there is an urgent need for both (i) greater application of policies and incentives to promote the conservation and sustainable use of biodiversity and ecosystem services, and (ii) more efficient use of available finance in existing biodiversity programmes. Payments for Ecosystem Services (PES) are a flexible, incentive-based mechanism that has potential to deliver in both of these areas.

What are Payments for Ecosystem Services and what is their role in biodiversity conservation and sustainable use?

PES are agreements whereby a user or beneficiary of an ecosystem service provides payments to individuals or communities whose management decisions influence the provision of ecosystem services. More specifically, PES are defined as "a voluntary, conditional agreement between at least one 'seller' and one 'buyer' over a well defined environmental service – or a land use presumed to produce that service" (Wunder, 2005). Ecosystem service beneficiaries include downstream hydroelectric utilities that use clean water as an input for production, and companies that benefit from value added when they sell organic products. The payments compensate individuals, such as farmers, foresters, or fishermen, for the additional costs of biodiversity and ecosystem service conservation and sustainable use, over and above that which is required by any existing regulations. As PES are voluntary, incentive-based instruments, seeking out sites with higher value and lower costs, they can provide potentially large gains in cost effectiveness compared to indirect payments or other regulatory approaches used for environmental objectives (Alix-Garcia *et al.,* 2003; Engel *et al.,* 2008).

Interest in PES has been increasing rapidly over the past decade. There are today more than 300 programmes implemented worldwide (Blackman and Woodward, 2010), predominantly used to address biodiversity, watershed services, carbon sequestration and landscape beauty (Wunder, 2006). PES are estimated to channel over USD 6.53 billion annually by national programmes in China, Costa Rica, Mexico, the United Kingdom and the United States alone. There are many more PES programmes that have a more limited geographic scope, with numerous local scale programmes operating in the developed and developing world.

Despite the proliferation of PES programmes, a common-cited criticism is that they fail to realise their potential cost-effectiveness gains (Ferraro, 2008; Wunder, 2007). This is because PES programmes often make fixed uniform payments on a per hectare basis. Such payments would be cost-effective if the costs and benefits of biodiversity and ecosystem service provision were constant across geographic space. This is not typically the case however. Instead, biodiversity and ecosystem benefits tend to vary from one location to another. Moreover, individual landholders are likely to have different opportunity costs of ecosystem service provision. The greater the spatial variation in costs and benefits, the larger the potential cost-effectiveness gains are when PES programmes are designed to take these differences into account.

How can PES best be designed to channel limited finance in the most cost-effective manner?

There are three elements that vary spatially in the context of PES (Wunscher *et al.,* 2006):

- the benefits of ecosystem service provision;
- the risk of ecosystem service loss, and the potential to enhance its provision; and
- the opportunity costs associated with ecosystem service provision.

Appropriate PES design, whereby ecosystem service buyers target and differentiate payments to account for this spatial variability can significantly enhance cost effectiveness. Metrics and indicators, including environmental or biodiversity benefit indices, can be developed to identify areas where benefits are highest. Scoring or weighting methods can help to prioritise payments, in particular when multiple ecosystem services are being targeted and when there are inherent trade-offs in their provision. To ensure that any ecosystem services paid for are indeed additional to those that would have occurred under a business-as-usual (*i.e.* baseline) scenario, payments should only be made to ecosystem services that are at risk of loss, or to enhance their provision. To estimate the opportunity costs of ecosystem service provision, and differentiate payments accordingly, administrators can obtain information on variables that affect opportunity costs (called costly-to-fake signals) such as agricultural prices, or they can use inverse auctions. Inverse auctions require potential ecosystem service sellers to submit bids indicating the minimum payment they are willing to accept for the provision of an ecosystem service.

How can the use of inverse auctions contribute to enhanced cost-effectiveness of PES?

Inverse auctions are suitable when there are a large number of bidders, thus inducing competition for payments. They are an innovative way to reflect sellers' opportunity costs in PES programmes, and can help maximise the ecosystem service benefits purchasable for the finance available. Auctions are being increasingly used in both developed and developing countries. For example, they have been applied in PES programmes to protect old growth forests in Australia, conserve waterfowl in Canada, reduce soil erosion in Indonesia, and improve agri-environment practices and enhance wildlife habitat in the United States.

Inverse auctions can effectively deliver large cost-effectiveness gains. In Australia for example, the inverse auction mechanism applied in the Tasmanian Forest Conservation Fund programme resulted in a 52% cost-effectiveness gain (compared to a first-come-first-served approach to allocating PES contracts). Likewise in the United States, a local PES programme in the Conestoga watershed found that the use of inverse auctions resulted in a seven-fold increase in the reduction of phosphorus runoff per dollar spent compared to a fixed price approach (Selman *et al.,* 2008).

What are the potential sources of PES finance and how can finance for PES best be mobilised?

Finance for PES can be mobilised directly from the ecosystem service users themselves, or from third-parties acting on behalf of the beneficiaries, such as governments or institutions. Since biodiversity provides benefits at the local, regional and global scale, how finance for PES can best be mobilised may depend on the geographic scale of the ecosystem service benefits. For example, if the objective is to address the local public good benefits of ecosystem services (such as watershed services), sources of finance can be mobilised at the local level from the users directly. If the objective is to address regional and global public good benefits, the most appropriate source of finance may be via governments or institutions at the national and international level, respectively.

What are the key criteria that must be addressed in PES programme design to enhance environmental and cost effectiveness?

The environmental and cost-effectiveness of PES depend crucially on programme design and implementation. Twelve key criteria that are essential to enhance PES effectiveness are:

1. *Remove perverse incentives*: For a PES programme to produce clear and effective incentives any conflicting market distortions, such as environmentally-harmful subsidies, should be removed.

2. *Clearly define property rights*: The individual or community whose land use decisions affect the provision of ecosystem services must have clearly defined and enforceable property rights over the land in question. Otherwise, risks associated with, for example, illegal logging or land appropriation will undermine the ability of a landholder to provide the ecosystem service, rendering the PES ineffective.

3. *Clearly define PES goals and objectives*: Clear PES goals help to guide the design of the programme, enhance transparency and avoid *ad-hoc* political influence.

4. *Develop a robust monitoring and reporting framework*: Monitoring and reporting of biodiversity and ecosystem services is fundamental, enabling the assessment of PES programme performance, and allowing for improvements over time.

5. *Identify buyers and ensure sufficient and long-term sources of financing*: Whether the buyers of services are the beneficiaries themselves, or third-parties acting on behalf of the beneficiaries, the finance must be sufficient and sustainable to ensure that the objective of the PES programme can be achieved.

6. *Identify sellers and target ecosystem service benefits*: Accounting for spatial variation in ecosystem service benefits via economic valuation, benefit scoring, and mapping tools allows payments to be prioritised to

those areas that provide the highest benefits. If the total PES budget available is limited, this can substantially increase the cost-effectiveness of the programme, in comparison to say, allocating payments on a first-come first-served basis.

7. *Establish baselines and target payments to ecosystem services that are at risk of loss, or to enhance their provision*: A PES programme should only make payments for ecosystem services that are additional to the business-as-usual baseline (*i.e.* in the absence of the programme).

8. *Differentiate payments based on the opportunity costs of ecosystem service provision*: PES programmes that reflect ecosystem providers' opportunity costs via differentiated payments are able to achieve greater aggregate ecosystem service provision per unit cost.

9. *Consider bundling or layering multiple ecosystem services*: Joint provision of multiple services can provide opportunities to increase the benefits of the programme, while reducing transaction costs, especially if finance for multiple benefits is available. The potential synergies and trade-offs involved in joint ecosystem service provision need to be identified.

10. *Address leakage*: Leakage occurs when the provision of ecosystem services in one location increases pressures for conversion in another. If leakage risk is expected to be high, the scope of the monitoring and accounting framework may need to be expanded to enable assessment of the potential leakage so that appropriate measures can be introduced to address it.

11. *Ensure permanence*: Events such as forest fires or illegal logging may undermine the ability of a landholder to provide an ecosystem service as stipulated in a PES agreement. If these risks are high, this will impede the effective functioning of a PES market. Insurance mechanisms can be introduced to address this.

12. *Deliver performance-based payments and ensure adequate enforcement*: Ideally, payments should be ex-post, conditional on ecosystem service performance. When this is not feasible, effort-based payments (such as changes in management practices) are a second best alternative, provided that changes in ecosystem management practices will bring about the desired change in service provision. Sufficient disincentives to breaching the PES agreement must also be provided and enforced,

especially if payments are based on efforts rather than on actual ecosystem service delivery.

What lessons can existing PES programmes offer for international PES?

The criteria and insights derived for designing and implementing effective local and national PES programmes are also relevant for the establishment of international PES (IPES). Examples of existing IPES-like activities include afforestation and reforestation projects under the Clean Development Mechanism, and more broadly, bio-prospecting agreements. A new international mechanism, Reducing Emissions from Deforestation and forest Degradation (REDD-plus), is also currently being proposed to help address the global climate change challenge. Successful agreement on a future REDD-plus mechanism would represent a substantial and unprecedented development in the creation of an international mechanism to help internalise the carbon-related ecosystem services provided by forests, and offers the potential to capture large biodiversity co-benefits (Karousakis, 2009).

IPES are likely to involve the need for greater institutional capacity including at the international level, for example for verification and review. The key building blocks for cost-effective PES, such as appropriate methods for targeting ecosystem services, remain the same. For biodiversity, which provides local, regional and global public good benefits, there is a need to consider how international finance for biodiversity can be mobilised to complement existing local and national PES programmes that target biodiversity benefits. Similarly, further work is needed on how emerging international voluntary initiatives that target both carbon and biodiversity can be improved and scaled-up.

References

Alix-Garcia, J., A. de Janvry and E. Sadoulet (2003), "Targeting Payments for Environmental Services: The Role of Risk", *Agricultural and Resource Economics Update,* 7(4).

Blackman A, and R. Woodward (2010). *User Financing in a National Payments for Environmental Services Program: Costa Rican Hydropower*, Resources for the Future, Washington DC.

Ferraro, P. (2008), "Asymmetric information and contract design for payments for environmental services", *Ecological Economics*, Vol. 65.

Karousakis, K. (2009), *Promoting Biodiversity Co-benefits in REDD,* OECD Environment Working Paper Series No. 11, OECD, Paris.

Selman, M., S. Greenhalgh, M. Taylor, and J. Guiling (2008), "Paying for environmental performance: potential cost savings using a reverse auction in program sign-up", World Resources Institute, Policy Note No. 5, Washington DC.

Wunder, S. (2005), "Payments for environmental services: Some nuts and bolts", CIFOR Occasional Paper No. 42, Center for International Forestry Research, Bogor, Indonesia.

Wunder, S. (2006), "Are direct payments for environmental services spelling doom for sustainable forest management in the tropics?", *Ecology and Society*, Vol. 11, No. 2.

Wunder, S. (2007), "The Efficiency of Payments for Environmental Services in Tropical Conservation", *Conservation Biology*, Vol. 21, No. 1.

Wunscher, T., S. Engel and S. Wunder (2006), "Payments for environmental services in Costa Rica: increasing efficiency through spatial differentiation", *Quarterly Journal of International Agriculture*, Vol. 45, No. 4.

Paying for Biodiversity: Enhancing the Cost-Effectiveness of Payments for Ecosystem Services

Introduction

This chapter introduces the different components of biodiversity and ecosystem services, the benefits they provide to society, and the categories of economic value that are associated with them. The underlying drivers of biodiversity loss and degradation are described and estimates on the costs of inaction are presented, demonstrating the need for renewed policy efforts to address this global environmental challenge. The chapter proceeds to discuss the role of Payments for Ecosystem Services in promoting the conservation and sustainable use of biodiversity and ecosystem services, and how PES fits into the broader policy framework.

Biodiversity and associated ecosystem service policies aim to promote "the conservation of biological diversity, the sustainable use of its components, and the fair and equitable sharing of the benefits arising out of the utilisation of genetic resources" (CBD, 1992). Despite the significant economic, social and cultural values of biodiversity and associated ecosystem services, biodiversity worldwide is being lost, and in some areas at an accelerating rate. It is widely acknowledged that the 2010 biodiversity target, agreed in 2002 under the UN Convention on Biological Diversity, to significantly reduce the rate of biodiversity loss by 2010 has not been met. Moreover, without significant new policy actions, OECD projections to 2030 indicate continuing biodiversity loss, driven primarily by land use changes (*e.g.* conversion to agriculture and infrastructure), unsustainable use and exploitation of natural resources, invasive alien species, climate change and pollution (OECD, 2008a). Given these trends in biodiversity loss, there is an urgent need for both (i) greater application of policies and incentives to address biodiversity and ecosystem service conservation and sustainable use, and (ii) more efficient use of available finance in existing programmes.

Biodiversity and the drivers of loss

Biodiversity is the "variability among living organisms from all sources, including, *inter alia*, terrestrial, marine, and other aquatic ecosystems, and the ecological complexes of which they are part: this includes diversity within species, between species and of ecosystems" (CBD, 1992). Ecosystem services are the beneficial outcomes, for the natural environment or people, which result from ecosystem functions (*i.e.* the physical, chemical, and biological processes or attributes that contribute to the self-maintenance of an ecosystem). More specifically, these benefits arise from the regulating, supporting, provisioning and cultural services that biodiversity and ecosystems supply (Figure 0.1). Provisioning services are the products obtained from ecosystems such as food, fuel, fresh water, and genetic resources; regulating services are the benefits obtained from the regulation of ecosystem processes such as air quality and climate regulation, and water purification. Cultural services refer to the nonmaterial benefits people obtain from ecosystems through, for example, recreation and aesthetic experiences; while supporting services are those that are necessary for the production of all other ecosystem services. Their impacts are often indirect or occur over a long time period. Examples include nutrient and water cycling, and photosynthesis (MA, 2005). Together, these services provide critical life support functions, contributing to human health, wellbeing and economic growth.[1]

Figure 0.1. Four components of ecosystem services

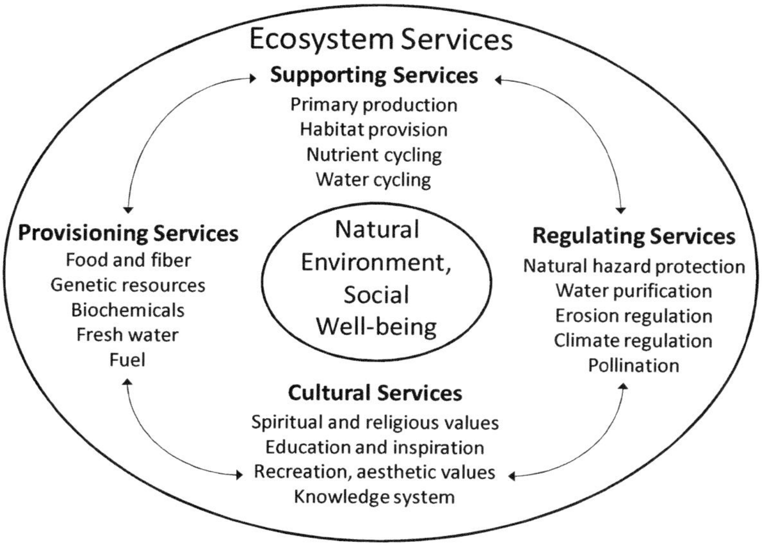

Source: OECD, 2010.

From an economic perspective, the aggregate benefits provided by biodiversity and ecosystems are comprised in the notion of Total Economic Value (TEV). TEV assesses the change in the values within each category of ecosystem services that occur as a result of changes in human activity (OECD, 2002). TEV aggregates both use and non-use values describing the different ways society values biodiversity and ecosystem services (Figure 0.2). Use values are derived directly from biodiversity in the form of consumables and indirectly through non-consumable services. Non-use values include existence values and bequest values, referring to the benefits individuals glean from the knowledge that biodiversity exists, and their wish to ensure it is passed on to future generations. Finally, option values reflect the value people place on the potential for future use, and how future advances in information can reveal new use and non-use values.

Figure 0.2. The Total Economic Value

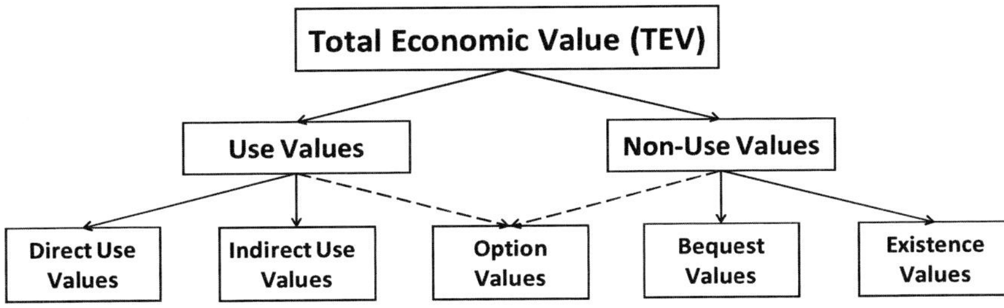

Source: OECD, 2010.

The total value of the benefits associated with biodiversity and ecosystem services is difficult to estimate however. The inherent ecological complexity and multidimensionality of biodiversity and ecosystems requires consideration of: uncertainty and imperfect information; thresholds and irreversibilities; the degree of substitutability between natural resources and other inputs; the treatment of the (very) long-run and distributional concerns; and, endogenous adaptation to changing conditions (OECD, 2002). Despite these difficulties in evaluating the total benefits of biodiversity and ecosystem services, studies suggest that they are very large. For example, it is estimated that the worldwide economic value of pollination services provided by insect pollinators (mainly bees), was USD 192 billion per year in 2005 for the main crops that feed the world (Gallai *et al.,* 2009). Similarly, the pharmaceutical industry relies on genetic diversity for drug developments; an estimated 25 to 50% of its business (about USD 650 billion per year) is derived from genetic resources (TEEB, 2008).

The total economic value of biodiversity provides a compelling case for investment in conservation and sustainable use. Current levels of financial flows for biodiversity conservation are estimated to be between USD 8 to 10 billion annually (James *et al.* 2001; Simpson 2004, Pearce 2007). The additional funding required to successfully conserve biodiversity depends on how the goals are defined. Some estimates suggest that an additional USD 19 billion annually may be needed to protect 70% of global biodiversity, through the acquisition of 2% of the Earth's terrestrial surface (Bruner *et al.,* 2004). Pursuing a more ambitious objective to establish a comprehensive system of reserves, protecting 10 to 15% of the world's surface, could cost an estimated half a trillion dollars (James *et al.,* 2001; Simpson 2004). These estimates move into the trillions of dollars if conservation on commercial forestry and agricultural land is included (James *et al.,* 2001; Simpson 2004).

While these figures may seem large, the costs of inaction in many areas are considerable. Estimates suggest that the aggregate loss of biodiversity ecosystem service benefits associated with the global loss of forests is between USD 2 and 5 trillion per year (TEEB, 2008). One UK study has estimated that the contribution marine biodiversity makes to climate regulation may be worth somewhere between USD 0.6 and 12.9 billion annually (Beaumont *et al.*, 2006). Just the collapse of the North Atlantic cod fishery and its closing in 1992 for example, resulted in short term costs of USD 235 million (*i.e.* the decline in landed value). In the long-term, the foregone potential annual income from a sustainable fishery was estimated at USD 0.94 billion per year (OECD, 2008). The total social costs of the fishery closure are even larger however as they extend beyond lost industry revenues to include other use and non-use values (Figure 0.3).

Figure 0.3. Costs of inaction with respect to biodiversity and ecosystem service loss

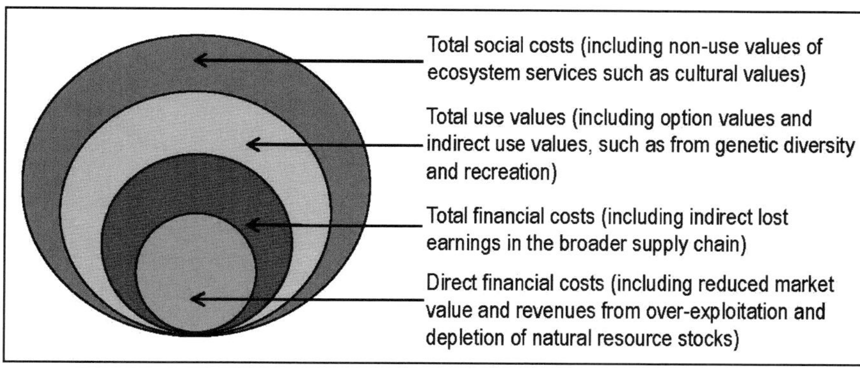

Source: Adapted from OECD, 2008.

The prevailing level of biodiversity and ecosystem service provision is below what would be socially optimal, due to market and policy failure. The optimal level is given by equating the benefit to society from conserving (or restoring) an additional unit of biodiversity, with the lost revenue from alternative land use or management decisions associated with conserving (or restoring) that unit.[2] The free market fails to achieve the optimal allocation because private decision makers do not consider the social costs and benefits of conservation, but rather consider only their own private costs and benefits. Market failure can be caused by the public good characteristics of biodiversity goods and services, the presence of externalities, imperfect information, and a lack of clear property rights.

Public goods are goods whereby consumption by one individual does not reduce availability of the good for consumption by others (non-rival); and whereby no one can be effectively excluded from using the good (non-excludable). As a consequence, there are limited incentives for individuals to invest in the provision of public goods, and everyone contributes too little, preferring the costs to be borne by their neighbours (referred to as free-riding). For example, the climate regulation services provided by forests are global public goods. The additional value of forests provided by their contribution to climate regulation generally is not considered in individuals' economic decision making, resulting in sub-optimal forest conservation. Other ecosystem services are quasi-public goods, being either excludable or rival, such as parks (excludable, non-rival), or fish stocks (non-excludable, rival).

Externalities occur when activities have a negative (or positive) impact on a third party, and when the resulting welfare loss is not compensated for. For example, a negative externality can occur when industrial water pollution imposes costs on downstream agricultural farmers without compensation for the loss in revenue, *i.e.* the costs have not been internalised.

Imperfect information can lead to market failure when individuals do not have complete knowledge of how biodiversity influences economic activity. For example, the services provided by wetlands in terms of hurricane protection and water filtration are only partially reflected in insurance schemes and drinking water markets, respectively, because individuals and firms may not be aware of the total benefits the services provide. The lack of recognition of the total value of these services results in under-investment to maintain their provision.

Clear and enforceable property rights provide individuals or communities with the authority to determine how a resource is used. Without distinct ownership or use rights, the good or service is effectively openly available to everyone. As in the case of open access fisheries for example, this can lead to over-exploitation as fishermen try to catch as much as possible, without taking into account the longer-term consequences of depletion.

Payments for Ecosystem Services and their role in biodiversity conservation and sustainable use

Payments for Ecosystem Services (PES) programmes aim to address market failure by providing direct incentives to enhance the provision of ecosystem services. PES compensate individuals or communities whose land use or other resource management decisions influence the provision of ecosystem services for the additional costs of providing these services.[3] More specifically, PES are defined as "a voluntary, conditional agreement between at least one 'seller' and one 'buyer' over a well defined environmental service – or a land use presumed to produce that service" (Wunder, 2007). Such payments are needed to help address the externalities associated with biodiversity and ecosystem services and the fact that they often display public good characteristics. PES are financed by the users of ecosystem services directly, or by third-parties such as governments or organisations acting on their behalf. In cases where the ecosystem services are public goods however, such as biodiversity, the incentives to free-ride may preclude the establishment of direct user-financed PES programmes. Moreover, as biodiversity provides local, regional and global public good benefits, the transaction costs associated with bringing together individual buyers and sellers can often be prohibitively high. In these circumstances, governments therefore often have an especially important role to play in facilitating PES programmes.

PES are based on a system where the user or beneficiary pays for the ecosystem services they would like to benefit from. This is in contrast to a system whereby the polluter is required to pay for the external environmental costs of their actions. The choice of instrument reflects the overall policy approach to the sector, the nature of property rights related to the use of natural resources (such as land and water) and societal and distributional concerns related to environmental issues (Vojtech, 2010). PES are one tool available to decision-makers for achieving positive environmental outcomes. They are flexible, incentive-based economic instruments which can be used alone or as part of a policy mix in conjunction with other instruments (Table 0.1). For example, PES can be used to incentivise enhancements in the provision of ecosystem services over and above that required by existing command and control regulation.

Ecosystem service payments are made directly to those who influence the provision of ecosystem services and as such have potentially large cost-effectiveness gains compared with other indirect and regulatory approaches (Alix-Garcia *et al.*, 2003; Engel *et al.*, 2008). This is because command and control approaches tend to impose uniform restrictions across landholders, requiring the same level of conservation from all. A PES approach is more flexible because participation is voluntary – landholders with relatively higher marginal costs of conservation will therefore tend to conserve less than those with lower costs. Moreover, indirect mechanisms proposed in the 1980s and 1990s to engage the development community and the private sector in biodiversity conservation – such as Integrated Conservation and Development Projects (ICDPs), sustainable product certification, ecotourism, and bioprospecting – tend to preserve biodiversity as a joint output used in the production of other goods and services. While these initiatives can work in certain circumstances, their relative lack of success at halting widespread species loss suggests that new mechanisms to harness conservation financing might best be both incentive-based and direct.

Table 0.1. **Policy approaches and instruments for biodiversity conservation and sustainable use**

Regulatory (*i.e.* Command and Control) Approaches	**Economic Instruments**	**Information and Other Instruments**
Restrictions or prohibitions on use (*e.g.* trade in endangered species and CITES) Restrictions or prohibitions to access (*e.g.* protected areas, legislated buffer zones along waterways) Quality or quantity standards, often enforcing the use of specific technologies (*e.g.* commercial fishing net specifications)	Price-based instruments • Taxes • Charges/Fees • Subsidies Liability instruments • Non-compliance fines • Criminal indictment • Performance bonds Removal or reform of perverse subsidies Market creation and assignment of well-defined property rights Payments for Ecosystem Services	Eco-labelling and certification Voluntary agreements

Source: OECD, 2010 [based on OECD (2008a) and OECD (2008b)].

Though the term PES is fairly new, "PES-like" instruments exist in a number of countries. Examples include agri-environmental programmes that are implemented across Europe to reduce the environmental impacts of intensive agriculture. PES-like vessel buyback schemes, such as those implemented in the United States Salmon fisheries since the 1970s, have also been used with the aim of reducing pressure on fishery stocks via diminished capacity (US GAO, 2001).

PES programmes are now being increasingly applied across developed and developing countries. There are today more than 300 PES programmes implemented worldwide (Blackman and Woodward, 2010), most of which have been set up to promote biodiversity, watershed services[4], carbon and landscape beauty (Wunder, 2006). It is estimated that over USD 6.53 billion is channelled by national PES programmes in China, Costa Rica, Mexico, the United Kingdom and the United States alone. To put this in context, in 2007 the OECD Development Assistance Committee (DAC) members allocated approximately USD 3.5 billion in bilateral Official Development Assistance (ODA) to biodiversity-related activities[5] (OECD, 2009); total global annual spending on protected areas is estimated at USD 6.5 billion (World Bank, 2006). Table 0.2 summarises the data on annual PES budgets across a selection of national and regional PES programmes. Most PES programmes have a more limited geographic scope, with numerous local scale programmes operating in the developed and developing world. Moreover, the global PES market is estimated to be increasing by 10 to 20% a year (Ecosystem Marketplace, 2008).

Table 0.2. **Annual PES budgets in selected national and regional PES programmes**

National PES Programmes	Annual Budget in USD
China, Sloping Land Conversion Programme (SLCP)	4 billion (Bennett, 2008)
Costa Rica, Payments for Environmental Services (PES)	12.7 million (FONAFIFO, 2009)
Mexico, Payments for Environmental Hydrological Services (PEHS)	18.2 million (Muñoz Piña *et al.*, 2008)
UK, Rural Development Programme for England	0.8 billion (Defra, 2009)
US, Conservation Reserve Program (CRP)	1.7 billion (Claassen, 2009)
Regional PES Programmes	**Annual Budget in USD**
Australia, Tasmanian Forest Conservation Fund (FCF)	14 million (DAFF, 2007)
Australia, Victoria State ecoMarkets	4 million (DSE, 2009)
Bulgaria and Romania, Danube Basin	575 000 (GEF, 2009)
Ecuador, Profafor	150 000 (Wunder and Alban, 2008)
Tanzania, Eastern Arc Mountains	400 000 (EAMCEF, 2007)

Source: OECD, 2010.

Despite the proliferation of PES programmes in the past decade, an often cited criticism is that they fail to realise their potential cost-effectiveness gains (Ferraro, 2008; Wunder, 2007). The environmental and cost-effectiveness of PES depends crucially on programme design and implementation. In practice, PES programmes differ in the type and scale of ecosystem service targeted, the payment source, the type of activity paid for, the performance measure used, as well as the payment mode and amount (Engel *et al.,* 2008).

Objectives, scope and structure of the book

This book aims to identify good practice in the design and implementation of PES programmes to understand how best to enhance their cost-effectiveness. The audience is policy makers at local, national, and international level. This book also responds to a call for further work in this area by the Convention on Biological Diversity (CBD) at its 9th Conference of Parties (COP-9) in 2008.[6]

The book draws on the literature on effective PES and on experience across more than 30 case studies in both developed and developing countries to make good practice insights accessible to policy practitioners. The following questions are addressed:

- Why are PES useful and how do they work?
- What are the key features that must be addressed in PES programme design to maximise their environmental effectiveness?
- How can PES best be targeted to channel limited finance most cost-effectively?
- How can the use of inverse auctions contribute to enhancing cost effectiveness?
- What are the different potential sources of finance for PES programmes, and how can they be secured?
- What are the lessons learned from existing PES programmes and insights for current and future programmes, including international PES?

The book is divided into two parts. Part I focuses on key issues for enhancing PES cost-effectiveness. Chapter 1 introduces the main concepts in the economics of PES. In Chapter 2, general pre-requisites and design issues for effective PES programmes are identified. Chapter 3 examines how to allocate ecosystem service payments in a more cost-effective manner. Methods and tools, such as environmental or biodiversity benefit indices and spatial mapping, to target ecosystem services with high benefits, high risk of loss and those where opportunity costs are low are reviewed. Chapter 4 presents options and experience with mobilising finance for PES. The relative merits of user and third-party (*e.g.* government) financing are discussed, as are innovative approaches to engaging and leveraging private sector finance. Chapter 5 considers the implications for international PES programmes. Part II proceeds to examine three PES case studies in depth. These are the US Conservation Reserve Program (CRP) (Chapter 6); the Tasmanian Forest Conservation Fund (FCF) in Australia (Chapter 7); and a pilot PES auction implemented in the Sumberjaya Watershed in Indonesia (Chapter 8). Together these case studies provide further insights on some of the challenges and lessons from PES applications that aim to target the costs and benefits of ecosystem service provision so as to enhance the environmental and cost-effectiveness of the programmes. A common element across the case studies is that they have applied an innovative feature, namely inverse auctions, to help achieve aggregate ecosystem service benefits at lower per unit cost. The US CRP is the longest running PES programme utilising inverse auctions. The FCF is a more recently designed inverse auction. The case study in Indonesia is one of the first applications of PES inverse auctions in a developing country. Finally, the conclusions chapter summarises the key policy-relevant outcomes and lessons learned to enhance the cost-effectiveness of current and future PES programmes.

Notes

1. For further discussion on the definition of ecosystem services, see also Brown *et al.,* 2007.
2. These are the opportunity costs.
3. PES are also applicable to aquatic and marine environments.
4. For PES recommendations related specifically to integrated water resources management, see Vermont *et al.,* 2007.
5. The OECD Development Assistance Committee is comprised of 24 member states, working on issues surrounding aid, development and poverty reduction (www.oecd.org/dac). Biodiversity-related aid is defined as activities that promote at least one of the three objectives of the UN Convention on Biological Diversity. These are: (i) the conservation of biodiversity, (ii) sustainable use of its components (ecosystems, species or genetic resources), and (iii) fair and equitable sharing of the benefits of the utilisation of genetic resources.
6. Decision IX/6 on Incentive Measures – Article 11.

References

Alix-Garcia, J., A. de Janvry and E. Sadoulet (2003), "Targeting Payments for Environmental Services: The Role of Risk", *Agricultural and Resource Economics Update*, 7(4).

Beaumont N., M. Townsend, S. Mangi, and M.C. Austen (2006), *Marine Biodiveristy: An Economic Valuation,* DEFRA, London.

Bennett, M. (2008), "China's sloping land conversion program: Institutional innovation or business as usual?", *Ecological Economics*, Vol. 65.

Blackman, A, and R. Woodward (2010), "User Financing in a National Payments for Environmental Services Program: Costa Rican Hydropower", *Resources for the Future,* Washington DC.

Brown, T, J. Bergstrom and S. Loomis (2007), "Defining, Valuing and Providing Ecosystem Goods and Services", *Natural Resources Journal*, Vol. 47.

Bruner, A.G., R.E. Gullison and A. Balmford (2004), "Financial needs for comprehensive, functional protected area systems in developing countries", *BioScience* Vol. 54.

CBD (Convention on Biological Diversity) (1992), Convention on Biological Diversity, CBD, Rio de Janeiro.

Claassen R. (2009), USDA Briefing Room – Conservation Policy: Background. http://www.ers.usda.gov/briefing/conservationpolicy/background.htm

DAFF, (Department of Agriculture, Fisheries and Forestry) (2007), Australia Sustainable Forest Management, DAFF, Canberra.

Defra (Department for Environment Food and Rural Affairs, UK) (2009), ERDP Schemes, http://www.defra.gov.uk/rural/rdpe/erdp/schemes/index.htm, accessed on 4th April 2010, London.

DSE (Deparment of Sustainability and Environment, Victoria, Australia) (2009), ecoMarkets, http://www.dse.vic.gov.au, accessed on 4th April 2010, Australia.

EAMCEF (Eastern Arch Mountains Conservation Endowment Fund) (2007), Welcome to the Eastern Arc Homepage, http://www.easternarc.or.tz/, accessed on 4th April 2010, Morogoro, Tanzania.

Ecosystem Marketplace (2008), *Mapping Ecosystem Service Markets, The Matrix,* The Ecosystem Marketplace and Forest Trends.

Engel, S., S. Pagiola and S. Wunder (2008), "Designing payments for environmental services in theory and practice: An overview of the issues", *Ecological Economics*, Vol. 65.

Ferraro, P. (2008), "Asymmetric information and contract design for payments for environmental services", *Ecological Economics*, Vol. 65.

FONAFIFO (*Fondo Nacional de Financiamiento Forestal* - National Fund for Forestry Financing in Costa Rica) (2009), Budget allocation for CAF and PSA by source of funding for the period 1995-2009 in Costa Rican Colonies, as of 19th March 2009.

Gallai, N., J.-M. Salles, J. Settele and B.E. Vaissière, (2009) "Economic valuation of the vulnerability of world agriculture confronted with pollinator decline", *Ecological Economics*, Vol. 68.

GEF (Global Environment Fund) (2009), Request for CEO Endorsement/Approval. GEF Washington DC.

James, A., K.J. Gaston, and A. Balmford, (2001), "Can we afford to conserve biodiversity?" *BioScience*, Vol. 51.

Millennium Ecosystem Assessment (MA) (2005), *Ecosystems and Human Well-Being: Synthesis*, Island Press, Washington DC.

Muñoz Piña, C., A. Guevara, J. Torres and J. Brana (2008), "Paying for the hydrological services of Mexico's forests: analysis, negotiations and results", *Ecological Economics*, Vol. 65(4).

OECD (Organisation for Economic Co-operation and Development) (2002), *Handbook of Biodiversity Valuation, A Guide for Policy Makers*, OECD, Paris.

OECD (2008a), *OECD Environmental Outlook to 2030*, OECD, Paris.

OECD (2008b), "Report in Implementation of the 2004 Council Recommendation on the Use of Economic Instruments in Promoting the Conservation and Sustainable Use of Biodiversity" [ENV/EPOC/GSP/BIO(2008)1/FINAL], OECD, Paris.

Pearce, D. (2007), "Do we really care about Biodiversity", *Environmental Resource Economics*, Vol. 37.

TEEB (2008), *The Economics of Ecosystems and Biodiversity*, Interim Report, TEEB, Bonn.

Simpson, R. D. (2004), "Conserving Biodiversity through Markets: A Better Approach", PERC (Property and Environment Research Center) policy series PS-32, United States.

U.S. General Accounting Office (GAO) (2001), "Effectiveness of fishing buyback programs can be improved", U.S. Testimony, GAO-01-699T U.S. GAO.

Vermont *et al.* (2007), *UN Recommendations on Payments for Ecosystem Services in Integrated Water Resources Management* http://www.unece.org/env/water/publications/documents/PES_Recommendations_web.pdf.

Vojtech, V. (2010), "Policy Measures Addressing Agri-environmental Issues", *OECD Food, Agriculture and Fisheries Working Papers*, No. 24, OECD Publishing. doi: 10.1787/5kmjrzg08vvb-en, Paris.

World Bank (2006), *Managing National Parks, How Public-Private Partnerships can Aid Conservation*, Public Policy for the Private Sector - Note Number 309, Washington DC.

Wunder, S. (2006), "Are direct payments for environmental services spelling doom for sustainable forest management in the tropics?", *Ecology and Society*, Vol. 11, No. 2.

Wunder, S. (2007), "The Efficiency of Payments for Environmental Services in Tropical Conservation", *Conservation Biology*, Vol. 21, No. 1.

Part I

Designing and implementing effective payments for ecosystem services programmes

Chapter 1

The economics of payments for ecosystem services

This chapter presents the main concepts in the economics of Payments for Ecosystem Services. The underlying mechanism for making payments for the provision of biodiversity and ecosystem services is illustrated in the context of market failures. The chapter also discusses how the use of spatially-explicit cost benefit analysis can help target the payments to enhance the cost-effectiveness of PES programmes.

PES programmes help address market failures by translating external non-market benefits of the environment into tangible financial incentives. Individuals or communities whose land use decisions affect the provision of ecosystem services are incentivised, via direct payments, to change their behaviour so as to reduce ecosystem service loss, or enhance their provision.

1.1 PES: an incentive-based mechanism

A PES can be illustrated through an example of a landholder's decision between two land use options: forest conservation and conversion to pasture (Figure 1.1). The landholder has greater potential (net) private benefits through conversion. In this example, this land use option however incurs costs to downstream ecosystem users or beneficiaries in the form of reduced watershed services, biodiversity loss, and carbon emissions. The 'minimum payment' that the landholder will be willing-to-accept as compensation to conserve the forest is the foregone opportunity cost of the alternative land use. The 'maximum payment' the ecosystem service beneficiary is willing-to-pay for conservation is the total costs of damage incurred when the land is converted to pasture. Thus, as is the case in Figure 1.1, if the potential benefits of conservation are larger than the minimum payment there is the potential for a mutually beneficial PES programme.[1]

Figure 1.1. The PES mechanism: ecosystem beneficiaries pay the landholder supplying the service to compensate for the additional costs of conservation

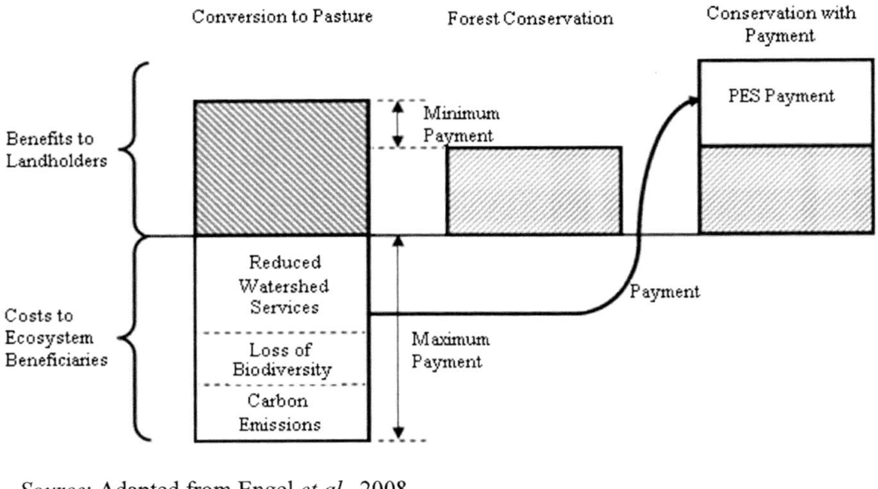

Source: Adapted from Engel *et al.*, 2008.

Participation in PES programmes is voluntary; rational landholders will enter into PES agreements as long as the payments cover at least their opportunity costs of changing their land use practices. Thus any payment level between the minimum and the maximum should be sufficient to induce a change in land use towards greater ecosystem service provision. Selecting the payment between these two levels has distributional and cost-effectiveness implications, but will bring about the same environmental change. This can be illustrated by looking at marginal costs and benefits.

Figure 1.2. Optimal provision of biodiversity, and the distributional and cost-effectiveness implications given a budget constraint

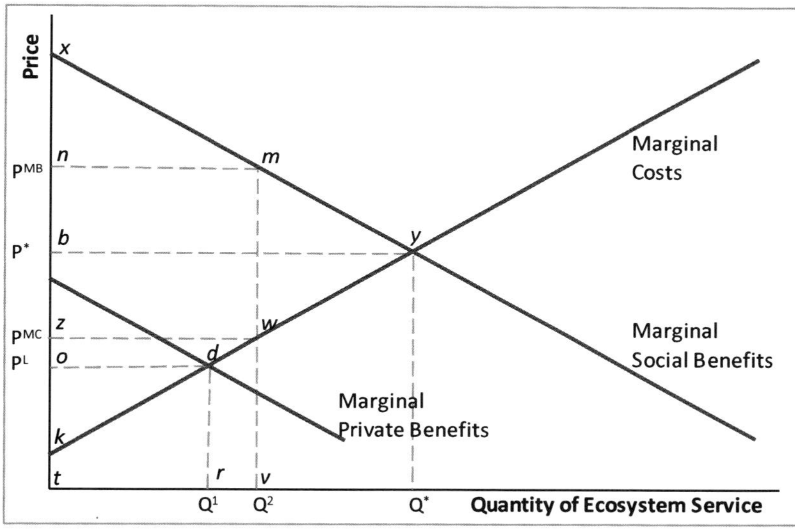

Source: OECD, 2010.

In Figure 1.2 the optimal provision of ecosystem services is given by Q^*, where the marginal costs of service provision are equal to the marginal social benefits. The costs of service provision include the opportunity costs of the alternative land use incurred by the landholder, and the transaction costs associated with the programme. The consumer surplus is given by area *yxb* and the producer surplus by area *ybk*. Due to the presence of market failures and the resulting divergence between private and social marginal benefits however, the prevailing provision of ecosystem services is Q^1 (*i.e.* in the absence of payment). To correct for market failure and achieve the socially optimal level of ecosystem service provision, a payment of P^* is necessary. In practice however, total offered payments may be insufficient to attain Q^* either because there are incentives for beneficiaries to free-ride, or because finance available (*e.g.* from government) for biodiversity and ecosystem service conservation and sustainable use is limited (Engel *et al.,* 2008). PES represent an improvement above the *status quo*.

At Q^2 for example, the marginal social benefits are greater than the marginal costs of provision. To achieve this level of service provision the payment level can be set between the two prices, P^{MB} and P^{MC}. These are analogous to the 'maximum' and 'minimum' payment level in Figure 1.1. Setting the payment according to the marginal costs, P^{MC}, allocates the greatest welfare surplus to the buyer, represented by area *wmxz*. Area *wzk* represents the private landholder's surplus. Conversely, setting the payment equal to the marginal social benefits, P^{MB}, allocates the greater welfare surplus, *wmnk*, to the private landholder, with *mxn* allocated to the buyer.

To maximise cost effectiveness, the minimum payment should be set, *i.e.* equal to the landholders marginal costs of service provision. Setting the price at P^{MC} requires a budget given by the rectangle *vwzt*, less *rdot.*[2] Conversely, setting the payment equal to the marginal social benefits, P^{MB}, the maximum payment, requires *vwzt* and *wmnz*, less *rdot*. Cost effectiveness, in terms of maximising the benefits from a given budget, therefore increases as the price moves towards P^{MC}.

In the context of biodiversity and ecosystem services, the levels P^{MC} and P^{MB} are likely to vary from one site to another depending on the magnitude of the ecosystem service benefits provided and the different costs incurred by landholders in their provision. To account for this spatial heterogeneity, spatially explicit cost benefit analysis is needed.

1.2 Spatial variability in the costs and benefits of biodiversity and ecosystem service provision

There are three elements that vary spatially in the context of biodiversity and ecosystem service provisioning (Wunscher *et al.,* 2006):

- the benefits of ecosystem services;
- the risk of biodiversity and ecosystem service loss, and the potential to enhance its provision; and
- the opportunity costs of their provision.

When the total supply of ecosystem services that landholders are willing to provide exceeds the available finance for the PES programme, the ability of a PES to maximise the total quantity of ecosystem services given the limited budget will depend on how buyers target and differentiate payments to selected landholders who can provide the maximum additional benefit per unit cost in a spatially explicit manner.

As indicated in the Introduction, the benefits of biodiversity and ecosystem services can be identified by estimating the different components of total economic value. Different valuation methods are available depending on components of value to be estimated (for a discussion on biodiversity valuation, see OECD, 2002). The risk or threat of ecosystem service loss can be estimated through an assessment of the business-as-usual scenario and an analysis of factors affecting land use changes. Identifying the opportunity costs incurred by the landholder in service provision can be achieved by gathering information on so-called costly-to-fake signals. Costly-to-fake signals refer to information that is correlated with opportunity costs, but is expensive or difficult for the landholder to artificially produce. For example, soil type can be used to infer the opportunity costs of agricultural land retirement through available information on productivity and crop prices. These techniques are needed because ecosystem service buyers are unaware of the costs of service provision incurred by landholders (*i.e.* the problem of information asymmetry). Moreover, landholders have an incentive to over-report their true opportunity costs so as to extract higher payments (and thus obtain larger producer surplus, or economic rent). Buyers are thus unable to select the lowest-cost providers.

Another approach to obtain information on opportunity costs is to use inverse auctions.[3] Inverse auctions require landholders to submit bids specifying the minimum amount they are willing to accept (WTA) as a compensatory payment for forgoing income from alternative land uses. The bids providing the highest ecosystem service benefits per unit cost are accepted until the budget is exhausted. Inverse auctions are most effective when (i) there is a large pool of potential suppliers, and (ii) if opportunity costs and service quality are considered to be heterogeneous across the potential service providers (Ferraro, 2008).

The competitive nature of auctions reduces the ability of suppliers to exploit the information asymmetry. Bidders must trade-off the risk of losing the contract to a competitor with extracting higher payments, and therefore have an incentive to bid closer to their true minimum WTA. Auctions do not completely eliminate information rents; the extent to which they succeed will depend on the level of competition and the bidders' preferences for risk (Hailu and Schilizzi, 2004). The inverse auction must therefore be carefully designed to maximise competition. Box 1.1 provides an overview of these design considerations. Though inverse auctions tend to involve greater transaction costs (Ferraro, 2008), experience suggests that they can nevertheless offer substantial cost-effectiveness gains.

Box 1.1. **Design considerations in inverse auctions**

There are two basic types of inverse auction: uniform-price auctions and discriminatory-price auctions. Uniform-price auctions set the same payment level for all successful bidders, usually the lowest rejected offer price. Discriminatory-price auctions, in contrast, pay successful bidders their bid price.

The advantage of uniform-price auctions is that bidders do not have an incentive to bid above their minimum WTA; over-bidding risks failing to be awarded a contract at an attractive price. The disadvantage is that they have cost-effectiveness losses associated with paying each landholder the same payment level irrespective of their opportunity costs, such that low-cost landholders are over-paid relative to their minimum WTA (Ferraro, 2008).

Discriminatory-price auctions can reduce these cost-effectiveness losses because the payment level for each landholder is designed to reflect their individual opportunity costs. However, bidders do have an incentive to inflate their bid. Maintaining high levels of competition is therefore important (Latacz-Lohmann and Van der Hamsvoort, 1997; Latacz-Lohmann and Schilizzi, 2005).

Sufficient competition can be ensured by minimising participant certainty of being successful. This can be achieved with a large pool of participants and careful auction design. For example, the bidders' information on the buyer's preferences in terms of maximum acceptable price, and preferred contract specification (where bids vary in ecosystem service benefits) can be reduced by keeping price caps hidden, and changing the details of the Environmental Benefits Index between successive auctions (Cason *et al.,* 2004). Similarly bidder information on the characteristics of competitors' bids and bidder collusion can be reduced by using sealed-bids and only allowing a single bidding round (*i.e.* prohibiting the revision of bids). These design considerations need to be carefully evaluated, in some cases trading-off theoretical competitive gains with participant understanding and process transparency (Rolfe and Windle, 2006).

The choice between uniform-price auctions and discriminatory-price auctions therefore depends on the anticipated level of competition achievable; if competition can be maintained discriminatory-price auctions are generally considered more cost-effective, however if over-bidding is considered to be a problem, uniform-price auctions may be preferable (Latacz-Lohmann and Schilizzi, 2005).

Notes

1. PES programmes are also applicable, for example, to incentivise reforestation of abandoned pasture lands that were originally forested.
2. Q^1 is provided by existing private incentives, representing the baseline level of service provision. Thus, finance is only required to purchase additional ecosystem service benefits; the movement from Q^1 to Q^2.
3. Screening contracts can also be used in theory, but in practice are complicated; see Ferraro (2008).

References

Cason, T., L. Gangadharan and C. Duke (2004), "A laboratory study of auctions for reducing non-point source pollution", *Journal of Environmental Economics and Management*, Vol. 46.

Engel, S., S. Pagiola and S. Wunder (2008), "Designing payments for environmental services in theory and practice: An overview of the issues", *Ecological Economics*, Vol. 65.

Ferraro, P. (2008), "Asymmetric information and contract design for payments for environmental services", *Ecological Economics*, Vol. 65.

Hailu, A. and S. Schilizzi (2004), "Are auctions more efficient than fixed price schemes when bidders learn?", *Australian Journal of Management*, Vol. 29, No. 2.

Latacz-Lohmann, U. and C. Van der Hamsvoort (1997), "Auctioning conservation contracts: a theoretical analysis and an application", *American Journal of Agricultural Economics*, Vol. 79.

Latacz-Lohmann, U. and S. Schilizzi (2005), "Auctions for conservation contracts: a review of the theoretical and empirical literature", Report to the Scottish Executive Environment and Rural Affairs Department.

OECD (Organisation for Economic Co-operation and Development) (2002), *Handbook of Biodiversity Valuation, A Guide for Policy Makers*, OECD, Paris.

Rolf, J. and J. Windle (2006), "Using field experiments to explore the use of multiple bidding rounds in conservation auctions", International Association of Agricultural Economics, IAAE Discussion Paper No. 1.

Wunscher, T., S. Engel and S. Wunder (2006), "Payments for environmental services in Costa Rica: increasing efficiency through spatial differentiation", *Quarterly Journal of International Agriculture*, Vol. 45, No. 4.

Chapter 2

Environmentally effective payments for ecosystem services

This chapter considers key design elements that need to be considered for the establishment of an environmentally effective PES programme. This includes ensuring that the necessary pre-requisites are in place, such as clearly defined property rights, and other design parameters such as a robust monitoring framework, establishing a business-as-usual baseline, and addressing environmental risks such as leakage and lack of permanence.

PES programmes are flexible, incentive-based instruments that can be used for different environmental objectives and can be designed in a number of ways. They have been used to provide financing to secure ecosystem services as diverse as water quality in Sweden, water quantity in Kenya, landscape quality in the United Kingdon, and carbon sequestration in Ecuador. More specifically in the context of biodiversity, PES have been adopted, for example, in Cambodia to help conserve the White Shouldered Ibis, one of the rarest birds in the world (Hirschfeld, 2009), and to enhance habitat quality in the United States. Other PES programmes aim to address multiple objectives, such as the Payments for Environmental Hydrological Services (PEHS) *(Pago de Services Ambientales Hydrologicas)* in Mexico which has a goal of reducing deforestation and water scarcity. Effective PES design and implementation is dependent on the specific goals, priorities and context of the programme. As noted, in practice, PES programmes differ in the type and scale of ecosystem service targeted, the payment source, the type of activity paid for, the performance measure used, as well as the payment mode and amount (Engel *et al.,* 2008).

This chapter considers some of the general design issues that need to be addressed to ensure environmentally effective PES programmes. This includes ensuring that the necessary pre-requisites are in place, such as clearly defined property rights and removing perverse incentives, together with setting up the necessary monitoring, reporting and verification framework so as to identify baselines and address possible leakage and permanence issues, as well as to enable appropriate PES performance review and enforcement.

2.1 PES pre-requisites

A key pre-requisite for a well-functioning PES programme is that property rights are clearly defined and enforced. Landholders providing ecosystem services need to have certainty that any management practices they invest in today will result in compensation, without risk of land appropriation or illegal activities such as logging. Lack of clearly defined property rights present significant barriers to the development of PES. In Brazil for example, "land grabbing, insecure tenure, overlapping claims, and lacking information on private tenure constitute real medium-term impediments to PES" (Borner *et al.,* 2010).

The reform of property rights needs to carefully consider all the social, economic and environmental implications. The UN Commission on Legal Empowerment of the Poor identified four key issues in property rights reform[1] (McAuslan, 2007):

- Local priorities – local authorities, under the guidance of national governments, should work with local communities to allocate land, register land titles, and manage disputes. Incursion and illicit activities may persist regardless of the reforms, thus PES programmes should consider the need for long-term funds for land protection and legal support of individuals and communities (Wendland, 2008).
- The role of traditional land rights – the traditional structure of land rights can often be used as a framework for property rights reform, while considering and incorporating the broader principles of equity and fairness.
- Access to information, justice and training – local communities need to be provided with transparent information and advice to get involved in the reform process and challenge the decisions of officials they consider to be adverse to their interests.
- Gender – principles of equality should be promoted in reforms of property rights.

For participation in PES programmes, Salzman (2009) makes the useful distinction between *de jure* and *de facto* land titles. *De jure* title describes ownership of the land, while *de facto* recognises only the occupancy and the practices taking place on the land. In many cases individuals may occupy the land and have influence over the provisions of ecosystem services, without legal *de jure* ownership, such as in the case of squatters or common property lands.

For example, in Mexico between the 1930s and 1980s the traditional land use patterns were formalised into common property lands called *ejidos*. Each household head, within the *ejido* was granted shared ownership rights, with decision making via a voting system carried out through a legally recognised authority. Participation in the Mexican PEHS forest conservation programme is thus undertaken through the *ejido* authority with payments split between the household heads (Kosoy *et al.*, 2008). In Nepal, Community Forestry User Groups were set up in the 1980's, granting *de jure* land rights to improve land stewardship and reduce deforestation. The clarification of property rights paved the way for hydroelectric companies to participate in PES programmes, enabling the conservation of the upland areas and secure cost savings from reduced reservoir dredging activities (Huang and Upadhyaya, 2007).

A second issue to be considered prior to the introduction of a PES programme is the broader domestic policy context. In many cases the causes of biodiversity loss are the result, at least in part, of other market distortions prevalent in the economy. For PES to produce a clear price signal and to function effectively, it is essential that any prevalent perverse incentives, such as environmentally harmful subsidies, are removed. Such subsidies artificially increase the profits of damaging production or increase consumptive activities, exacerbating the market failure, and increase the opportunity cost of undertaking environmental activities. In Indonesia for example, Jack *et al.* (2007) note that the Rewarding Upland Poor for Environmental Services programme incentivises farmers to maintain mixed agro-forestry for rubber; the government, however, simultaneously provides subsidies to clear the forest and convert it to rubber monocultures. Similarly, in Mexico cattle ranching subsidies totalling USD 800 million per year effectively encourage deforestation, and are inconsistent with the aims of the PEHS programme (Muñoz Piña, 2010). These subsidies distort the true price signal, and counter-effect the incentives provided by ecosystem service payments. Policy coherence across different sectors in the economy is therefore needed.

To promote policy cohesion, the creation of a high-level governing board or steering committee comprising multiple stakeholders can help foster stakeholder involvement, enhance co-ordination and provide oversight to the PES programme. In the Costa Rican PES forestry conservation programme for example, a governing board was established to oversee the programme design and implementation strategy. This included officials from the Ministry of Environment and Energy, the Ministry of Agriculture and Livestock, the national banking system, as well as representatives from the private sector (Pagiola, 2006).[2]

2.2 General design elements for environmentally effective PES

Clear goals and objectives

PES programmes must firstly clearly set out their goals and objectives. This will help to guide the effective design of the programme. Experience with environmental funds for example, has shown that a lack of clearly defined goals can lead to larger numbers of grant-seeking proposals, and thus higher administrative and transaction costs, as well as delays in the disbursement of funds (Norris, 2000). This implies fewer resources available for activities or projects that directly benefit the environment and greater difficulty for the poor to access funds. Similarly for PES programmes, if the goals are ambiguous, the rules and resulting outcomes may diverge from desired objectives (at least for some participants).

Identifying clear PES goals and objectives requires an understanding of the current and projected magnitude of the biodiversity and ecosystem service problem that is being addressed, and the underlying socio-economic drivers of biodiversity and ecosystem service degradation and loss.

Monitoring, reporting and review

A robust monitoring and reporting framework is fundamental for an effective PES programme and allows for an assessment of whether the PES programme is delivering its intended objective. It therefore also enables decision-makers to adjust and improve PES programme design over time.

Monitoring should be undertaken at three levels: (i) the implementation level, to assess that landholders are undertaking the contracted land use; (ii) the ecosystem services level, to ensure that changes in land use are enhancing the provision of services; and (iii) at the participants level, to assess socio-economic impacts and ensure that welfare of participants is improved.

In the Costa Rican Payments for Environmental Services (*Pago de Services Ambientales*) for example, monitoring and reporting is conducted through various activities, including via Geographic Information System (GIS) and an Integrated Project Management System (IPMS). The IPMS is composed of several modules: general planning, procurement and contracts, financial administration, monitoring of physical progress, evaluation of results, and the PES programme (see Box 2.1).

Box 2.1. **The integrated project management system for the Costa Rican PES**

Contracts: Ensures that contracts and procurements for projects are implemented in a timely manner, to the expected standards, at reasonable prices and using efficient, effective and transparent processes.

Finance: Facilitates the efficient flow of project funds, in line with the Implementation Plans and with the requirements of the financiers.

Accounting: Generates useful information on the financial execution of the Projects.

Fixed Assets: Facilitating control of the assets procured.

Monitoring and Evaluation of Results: Facilitates the timely identification of achievements, variances, risks, weaknesses and corrective actions in the physical and financial execution of the Projects, to enhance their results.

Planning and Budgets: Facilitates the rational and timely preparation of plans and budgets for the execution, follow-up and quantitative evaluation of the physical and financial outputs of the projects.

Payments Environmental Services System: Facilitates the input of data relevant to the PES contracts, the processing of payments and the monitoring of the areas subject to the PES Programme.

The Mexican PEHS uses high resolution satellite imaging technology to monitor geographically dispersed forest areas. Participating lands are monitored once a year, together with some of the surrounding area in an effort to detect leakage (Muñoz Piña *et al.,* 2008). The initial development costs of monitoring were USD 5.6 per hectare, relative to payments of USD 30 per hectare (*i.e.* a ratio of about 1:5). In comparison, the on the ground monitoring used in the Pimpampiro PES programme in Ecuador has a lower monitoring cost to service payment ratio (1:8), however it is limited by personal capacity and budget constraints (Wunder and Alban, 2008). In three PES programmes implemented in Cambodia for biodiversity conservation, monitoring is conducted at the local level by village institutions, by an external agency for certification, and by the Protected Area management for the enforcement of national laws (Clements *et al.*, 2010).

The type of data and monitoring methods used to assess ecosystem service provision will need to be tailored to the environmental objective of the PES mechanism. Ideally, payments should be made directly on the basis of the measure of biodiversity or ecosystem service provided. For example, if the aim is to conserve old growth forests, data on deforestation and degradation of these species will be needed. If the aim of the programme is to conserve waterfowl, population growth and nesting success may provide the most appropriate data. There may be trade-offs involved between the accuracy of the monitoring methods used versus the costs of implementation. Moreover, many ecosystem services cannot be observed by the landholders. In the case of biodiversity for example, the impact of individual actions are hard to separate from those of their neighbours (Engel *et al.,* 2008). Proxies or indicators may need to be developed so as to reduce monitoring costs. Many PES programmes make payments on a per-hectare basis. Proxies that are too aggregated however can undermine the cost-effectiveness of the PES programme, an issue that is discussed in Chapter 3 on targeting.

Baselines and additionality

Baselines are an essential element of any mechanism aiming to address biodiversity and ecosystem service loss and degradation. They provide information on the expected trends in the provision of these services and hence the magnitude of the incentives that will be needed to attain a certain goal, as well a reference against which performance can eventually be assessed (discussed below). Baselines refer to the business-as-usual scenario of trends in ecosystem services in the absence of new policies. Historical trend data is a starting point but needs to be combined with projections of key variables, such as population and economic growth, to provide forward projections.

Baselines are critical to ensuring that any payment leads to additional benefits relative to the *status quo*. For example, payments for habitat protection are only additional if in their absence the habitat would be lost. Low additionality has been raised as an issue in several PES programmes, including Finland and Costa Rica, because of the low risk of imminent forest loss (Zandersen *et al.*, 2009; Wunscher *et al.*, 2006). Clear understanding of whether or not ecosystems are at risk of loss or degradation is therefore needed. Appropriate monitoring and reporting frameworks and the institutions to support this are required for this.

Baselines can also help to minimise problems of perverse incentives from 'new polluters' *i.e.* those who threaten to degrade ecosystems after a PES programme has been introduced so as to obtain payments. For example, after the introduction of a PES programme in Austria, some landholders threatened to start polluting to attract payments. Baselines can therefore help to set up-front eligibility criteria for participation in a PES programme and therefore enhance additionality. Eligibility criteria has been used, for example, in the US agri-environmental set-aside programmes: landholders must have cropped the land for several years prior to enrolment into the programme, and cannot have purchased the land for the purpose of enrolment (see Chapter 6). Similarly, the Scottish Challenge Fund afforestation scheme was deemed to attain high additionality through stringent eligibility criteria, and also because in the absence of the scheme there are negligible financial incentives for landholders to re-plant woodlands (Latacz-Lohmann and Schilizzi, 2005; Zandersen *et al.*, 2009).

Avoiding leakage

Leakage occurs when securing an ecosystem service in one location leads to increased pressure to convert or degrade ecosystem services in another. Leakage can occur at the intra-national or international level, but it is only an issue if changes in ecosystem service provision occur outside the monitoring and accounting framework. The extent to which risk of leakage is a concern depends on the price elasticity of supply and demand for ecosystem services. If risk of leakage is anticipated to be high, the monitoring framework may need to be extended beyond the geographic boundaries of the PES programme so as to assess the magnitude of leakage and measures introduced in the design of PES to address this.

To avoid intra-property leakage in the Mexican PEHS (which aims to mitigate deforestation and address water scarcity), in many cases the PES contracts specify that the removal of trees from the community's entire forest area (even outside of the area for which payments are being made) constitute a PES contract violation and hence subsequent non-payments. Other measures to mitigate leakage include introducing additional complementary economic incentives, such as increased taxes on converted native land. In the United States, the removal of agricultural subsidies to recently converted land has been suggested (see Chapter 6).

It is important to note that expanding the geographic scope of the monitoring and reporting framework is likely to raise the implementation costs of the programme. The expected risk and magnitude of leakage therefore needs to be balanced with the additional expenditure this will entail.

Permanence

Permanence refers to the ability to ensure the provision of the ecosystem service over the long-term. Ecosystem service payments provide the necessary incentives for landholders to change their land-use decisions or management practices. Once the payments cease, the landholder will no longer have the added incentive needed to provide a greater level of ecosystem services. This is one of the advantages of PES programmes, allowing flexibility and adjustments in PES to reflect changes in market conditions (such as agricultural food prices). PES agreements entail contracts of a specified length, at the end of which all involved can consider contract renewal. PES programmes should therefore entail continuous payments.

The long-term provision of biodiversity and ecosystem services may however be undermined by unforeseen events such as fires, hurricanes, and the invasion of alien species, or other human-induced occurrences such as illegal logging. The allocation of responsibility and risk therefore needs to be specified in the conservation contract. If these risks of non-permanence are particularly high, insurance payments, or the creation of an emergency rehabilitation fund, can be considered.

Typically, where the loss of service provision is directly or indirectly due to negligence on the part of the ecosystem service provider, payment can simply be withheld. In the Mexican PEHS for example, if there is purposeful breach of contract on behalf of the ecosystem service provider, then there is no payment at the end of the year, irrespective of how small the land-use change is.

Performance-based payments and enforcement

To successfully deliver the desired ecosystem service outcome, payments should be ex-post and conditional upon actual delivery of the ecosystem services themselves. In Sweden, for example, a waste water treatment plant makes direct payments to blue mussel farmers based on the measured nitrogen and phosphorus content of the harvested mussels' biomass (Zandersen *et al.*, 2009). In another Swedish programme, payments are made to reindeer herders in Sami Villages based on the reproductive success of large carnivores thus disincentivising poaching (Zabel and Holm-Muller, 2007).

In some cases however, performance based payments might not be feasible due to concerns such as the high costs of monitoring ecosystem services directly, or the time delay between the implementation of the management practice and the ecosystem service provision (see Indonesian case study in Chapter 8). Under these circumstances, an alternative is to use proxy-based payments. In China for example, the Sloping Land Conversion Programme pays landholders for planting erosion protection cover according to surface area; the success of erosion reductions does not affect the payment (Bennett, 2008). In the US Conservation Reserve Programme, proxies used for wildlife habitat benefits are types of vegetative cover (USDA, 2006). Effort-based payments, whereby payments are made based on actions presumed to supply a given ecosystem service, are another form of a proxy-based payment. Examples of effort-based payments in the context of farm-level management include conservation tillage (to enhance carbon sequestration in soil), and changes in rice paddy management (to reduce methane emissions). Such payments are suitable as long as there is a strong relation between the management practices undertaken by the landholder and the resulting ecosystem service provided. Effort-based payments, however, can be subject to problems of moral hazard, especially where monitoring efforts is also costly and penalties for breach of contract are weak.

A robust monitoring and reporting framework facilitates effective enforcement of the PES programme and the application of non-compliance penalties and fees when necessary. Non-compliance in the Mexican PEHS is penalised by the withdrawal of current and future payments. In two years, three cases of non-compliance have been punished in this way (Wunder *et al.*, 2008; Muñoz Piña *et al.*, 2008). The Environmental Quality Incentives Program (EQIP), in the United States, may suffer from poor contract compliance with an estimated 17% non-compliant in some way (Cattaneo, 2003). The level of enforcement may not provide adequate disincentive for breaching contract because of a reluctance to penalise due to legal costs.

Notes

1. In relation to Africa, but the principles are transferable.
2. For further information on the legal, institutional, financing and environmental framework of the Costa Rican PES, see Karousakis, 2007.

References

Bennett, M. (2008), "China's sloping land conversion program: Institutional innovation or business as usual?", *Ecological Economics*, Vol. 65(4).

Borner, J., S. Wunder, S. Wertz-Kanounnikoff, M. Tito, L. Pereira, N. Nascimento (2010), "Direct conservation payments in the Brazilian Amazon: Scope and equity implications", *Ecological Economics*, Vol. 97, No. 6.

Cattaneo, A. (2003), "The pursuit of efficiency and its unintended consequences: contract, withdrawals in the environmental quality incentives program", *Review of Agricultural Economics*, Vol. 2.

Clements, T., J. Ashish, K. Nielsen, D. An, S. Tan, and E. Milner-Gulland (2010), "Payments for biodiversity conservation in the context of weak institutions: Comparison of three programs from Cambodia", *Ecological Economics*, Vol. 69.

Engel, S., S. Pagiola and S. Wunder (2008), "Designing payments for environmental services in theory and practice: An overview of the issues", *Ecological Economics*, Vol. 65.

Hirschfeld E. (2009), in E. Hirschfeld (ed.), *Rare Birds Yearbook 2009*, Birdlife International, Cambridge.

Huang, M. and S. Upadhyaya (2007), "Watershed-based payments for environmental services in Asia" Winrock International, Working Paper No. 06-07, Arkansas, United States.

Jack, B.K., K. Kousky and K. Sims (2007), "Designing payments for ecosystem service: Lessons from previous experience with incentive-based mechanisms", *Proceedings of the National Academy of Sciences,* PNAS, Vol. 105, No. 28.

Karousakis, K. (2007), "Incentives to Reduce GHG Emissions from Deforestation: Lessons Learned from Costa Rica and Mexico", OECD, Paris.

Kosoy, N., E. Corbera and K. Brown (2009), "Participation in payments for ecosystem services: Case studies from the Lacandon rainforest, Mexico", *Geoforum,* Vol. 30, No. 6.

Latacz-Lohmann, U. and S. Schilizzi (2005), "Auctions for conservation contracts: a review of the theoretical and empirical literature", Report to the Scottish Executive Environment and Rural Affairs Department.

McAuslan, P. (2007), "Improving tenure security for the poor in Africa", synthesis report, UNFAO, Washington DC.

Muñoz Piña, C. (2010), "Measuring the REDD effect of the Payments for Environmental Hydrological Services Program in Mexico (PEHS)", Presentation given to OECD Workshop 25th March 2010. www.oecd.org/env.biodiversity.

Muñoz Piña, C., A. Guevara, J. Torres and J. Brana (2008), "Paying for the hydrological services of Mexico's forests: analysis, negotiations and results", *Ecological Economics*, Vol. 65.

Norris, R. (eds.) (2000), *The IPG Handbook on Environmental Funds: A resource book for design and operation of environmental funds*, The Interagency Planning Group, Washington DC.

Pagiola, S. (2006), "Payments for Environmental Services in Costa Rica", *Munich Personal RePEc Archive.*

Salzman, J. (2009), "A Policy Maker's Guide to Designing Payments for Ecosystem Services", *Duke Law Faculty Scholarship,* Paper 2081.

USDA (United States Department of Agriculture) (2006), Fact Sheet, Conservation Reserve Program General Sign-up 33, Environmental Benefit Index, USDA, FSA, Washington.

Wendland, K. (2008), "Rewards for ecosystem services and collective land tensure: lessons from Ecuador and Indonesia", Tenure Brief, Land Tenure Centre, Wisconsin.

Wunder, S. and M. Albán (2008), "Decentralized payments for environmental services: the cases of Pimampiro and PROFAFOR in Ecuador", *Ecological Economics*, Vol. 65, No. 4.

Wunder, S., S. Engel, and S. Pagiola (2008), "Taking stock: A comparative analysis of payments for environmental service programs in developed and developing countries", *Ecological Economics*, Vol. 65, No. 4.

Wunscher, T., S. Engel and S. Wunder (2006), "Payments for environmental services in Costa Rica: increasing efficiency through spatial differentiation", *Quarterly Journal of International Agriculture*, Vol. 45, No. 4.

Zabel, A. and K. Holm-Muller (2007), "Conservation performance payments for carnivore conservation in Sweden", *Conservation Biology*, Vol. 22, No. 2.

Zandersen, M., K. Braten, H. Lindhjem (2009), "Payment for and Management of Ecosystem Services, issues and options in the Nordic context", Nordic Council of Ministers, Copenhagen.

Paying for Biodiversity: Enhancing the Cost-Effectiveness of Payments for Ecosystem Services

Chapter 3

Cost-effective targeting of payments for ecosystem services

Individuals or communities with the potential to influence the supply of ecosystem services will often differ in the magnitude of benefits they can provide, the risk that these services will otherwise be lost or the extent to which their management activities can enhance biodiversity and ecosystems, as well as the costs of service provision. This chapter discusses how PES programmes can be designed to address these issues, and presents the tools and methods through which payments can be targeted to increase PES cost-effectiveness.

How payments for biodiversity and ecosystem services are targeted is critical in determining the cost-effectiveness of a PES programme. In most cases, the available budget for biodiversity and associated ecosystem services will be limited and competing with different demands. Cost-effective targeting of payments enables greater total benefits to be achieved with a given PES budget, and can therefore also contribute to the long-term success of the programme.

Many PES programmes allocate uniform payments on a per hectare basis. This is cost effective if ecosystem service benefits and the costs of their provision are constant across space. In many cases however, this is unlikely. The more heterogeneous the costs and benefits are, the greater the cost-effectiveness gains that can be realised via targeted and differentiated payments. Indeed, more and more PES programmes are incorporating design elements to address this. This chapter examines the methods and tools that are available to target spatial heterogeneity in biodiversity and ecosystem service benefits, the threat of loss, and the costs of their provision.

3.1 Targeting ecosystem services with high benefits

Identifying areas with high biodiversity and ecosystem service benefits requires metrics and indicators to quantify them. Selecting an appropriate metric or indicator for PES that aims to enhance biodiversity conservation and sustainable use is not necessarily straightforward however. Unlike carbon for example, which is measured in tCO_2e, there is no single standardised metric to quantify biodiversity. The multidimensionality and the inherent complexity of biodiversity requires trade-offs between the accuracy of a metric and the costs of development. The appropriate biodiversity metric or indicator selected for a PES programme may also depend on the specific objectives of the programme. Indeed, methodologies for constructing metrics and indicators tend to be tailored to specific local, regional and national programmes and their objectives. Examples of metrics and indicators used across two biodiversity PES programmes, namely the Victorian BushTender programme in Australia, and the PES implemented in the Assiniboine River watershed of east-central Saskatchewan programme in Canada are presented in Box 3.1.[1]

Box 3.1. Metrics and indicators used to target biodiversity benefits in the Victorian BushTender and a Canadian pilot PES

The Habitat Hectare Method in the Victorian BushTender Programme

The aim of Victorian BushTender programme in Australia is to improve the management of native vegetation on private land. To quantify biodiversity benefits, the BushTender programme uses the Habitat Hectare (HH) methodology. The HH is comprised of an assessment of the local benefits via the Biodiversity Benefits Index (BBI). The BBI is based on the proposed management practices; the conservation significance in terms of regional priorities through the Biodiversity Significance Score (BSS), the cost of conserving the land (b), and the size of the proposed land (ha). Potential plots are compared through an inverse auction, where landholders submit bids including information on the proposed area, the BBI, and the required payment. The BSS is calculated separately to improve competition (DSE, 2004).

$$HH = BBI \times ha$$

$$BBI = (BSS \times HSS)\, b$$

where HH = Habitat Hectare; BBI = Biodiversity Benefits Index; ha = area in hectares

BSS = Biodiversity Significance Score; HSS = Habitat Service Score; b = cost of bid

Targeting Waterfowl in a Canadian pilot PES programme

In Canada a pilot PES programme, initiated in 2008, to restore drained wetlands was undertaken in the Assiniboine River watershed of east-central Saskatchewan. The Environmental Benefits Index was based on the incremental increase in predicted hatched waterfowl nests relative to bid price. The EBI was based on the Ducks Unlimited Canada Waterfowl Productivity Model (DUC) which evaluated the potential of wetland restoration on each plot to increase the number of hatched waterfowl nests in the Assiniboine Watershed. The EBI was based on wetland area restored, waterfowl density, existing wetland density, and the percentage of cropland in a 4 x 4 mile block around the plot (Hill *et al.*, forthcoming).

The use of such metrics to better target ecosystem service payments can substantially enhance PES cost-effectiveness. In the Tasmanian Forest Conservation Fund programme for example, a comparison of using the AUD/CVI metric with a simpler AUD/ha metric indicated an 18.6% gain in conservation outcomes. Comparing the additional conservation gains (valued at approximately AUD 3.3 million) with the costs of achieving those benefits (AUD 0.5 million), illustrate that the ratio of benefits to costs from investing in the CVI is 6.9:1 (see Chapter 7). Similarly, Wunscher *et al.* (2006) simulated different targeting approaches for the Costa Rican PES and estimated that a scenario selecting highest scoring sites with the given budget would have resulted in 14% higher benefits than the current system of selecting sites (see Box 3.2).

Spatial mapping tools

Spatial mapping tools are increasingly being used to discern the spatial heterogeneity in ecosystem costs and benefits. Several of these tools are emerging to help design PES systems at the regional and national level, however there are increasingly initiatives of spatial mapping tools that are being developed at the international scale, including the UNEP-WCMC Carbon and Biodiversity Demonstration Atlas, ARtificial Intelligence for Ecosystem Services (ARIES),[2] the Integrated Valuation of Ecosystem Services and Trade-offs (InVEST)[3] and SENSOR.

Figure 3.1. Targeting PES in Madagascar

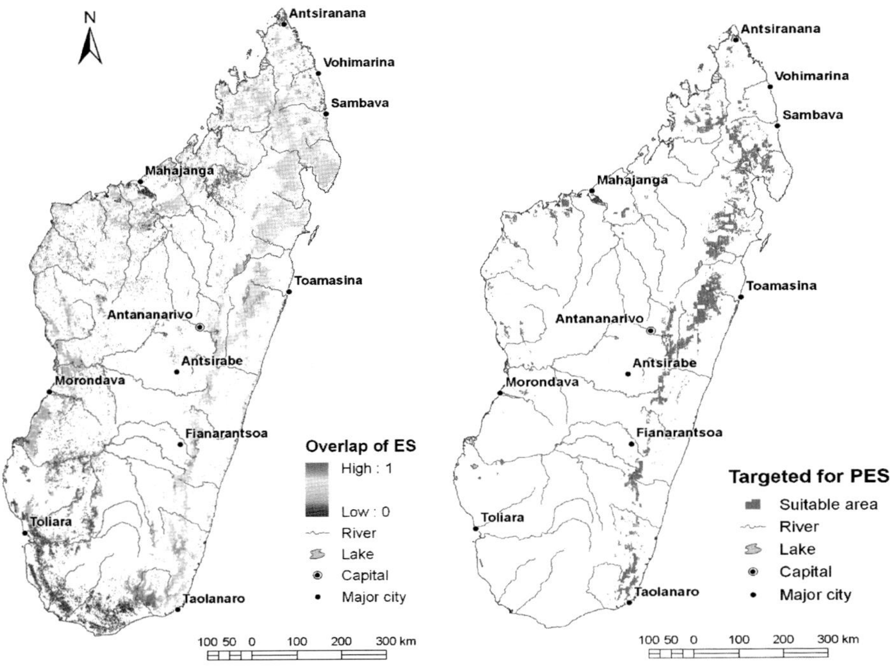

Source: Adapted from Wendland *et al.* 2009.

To target ecosystem service payments in Madagascar, Wendland *et al.* (2009) examined the spatial distribution of biodiversity (proxied by vector data on species ranges of mammals, birds and amphibians), carbon and water quality. The left panel of Figure 3.1 depicts the degree of overlap between these three ecosystem services. The right panel further incorporates information on the probability of deforestation and the opportunity cost of the land to identify where payments could be most cost-effectively targeted.

One example of a spatial mapping tool developed at the international level is the Carbon and Biodiversity Demonstration Atlas, produced by UNEP's World Conservation Monitoring Centre (UNEP-WCMC) (Kapos *et al.,* 2008). The Atlas includes regional maps as well as national maps for six tropical countries showing where areas of high biodiversity importance coincide with areas of high carbon storage. Figure 3.2 illustrates the national map for Panama, indicating that 20% of carbon is stored in high carbon, high biodiversity areas.

To identify areas of high biodiversity importance for the regional maps, UNEP-WCMC uses six indicators for biodiversity, namely Conservation Internationals' Hotspots, WWF 200 Ecoregions, Birdlife International Endemic Bird Areas, Amphibian Diversity Areas, Centers of Plant Diversity, and the Alliance for Zero Extinction Sites. Areas of high biodiversity, as determined by UNEP-WCMC, are areas where at least four of the above listed biodiversity-conservation priority areas overlap with areas in dark green indicating a greater degree of overlap.

Figure 3.2. Example of a UNEP-WCMC national map: Panama

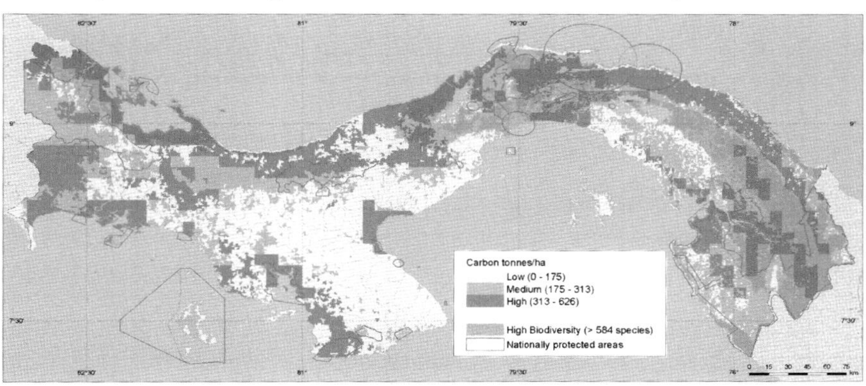

Source: Kapos *et al.,* 2008.

The maps identify the different areas with high biodiversity importance. The maps do not necessarily identify areas with high biodiversity benefits in economic terms. Ideally, spatial maps on biodiversity benefits would incorporate the total economic value of these sites, with an assessment of both direct and indirect use values.

A number of spatial mapping initiatives are currently underway and are in different stages of development. These include ARtificial Intelligence for Ecosystem Services (ARIES) (Villa *et al.,* 2009); InVest (Tallis *et al.,* 2010); the USGS Global Ecosystems initiative;[4] and SENSOR (Sustainability Impact Assessment: Tools for Environmental, Social and Economic Effects of Multifunctional Land Use in European Regions).[5]

Figure 3.3. Marketing biodiversity joint service provision

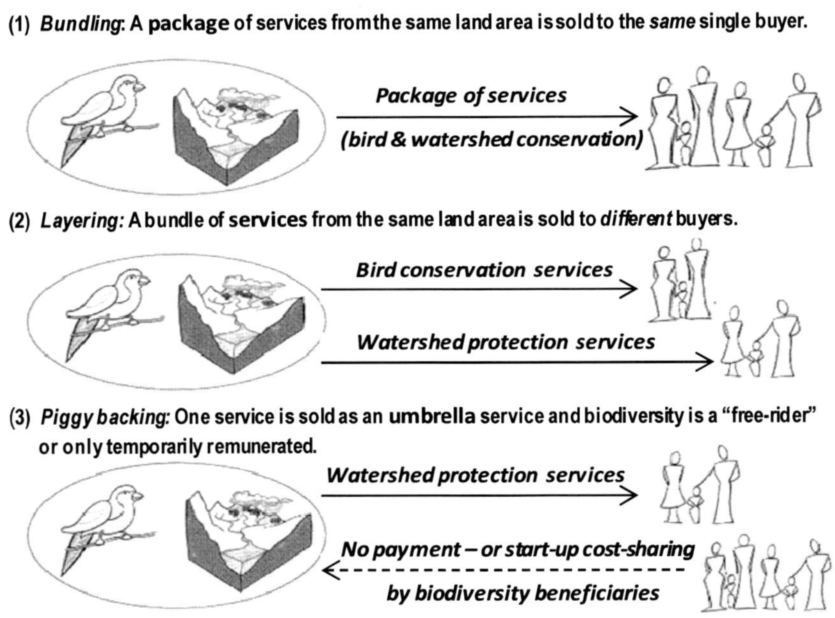

Source: Wunder and Wertz-Kannounikoff, 2009.

As suggested in the Madagascar example above, PES programmes can simultaneously target multiple ecosystem service benefits. Bundling or layering (see Figure 3.3) can allow a broader range of ecosystem service benefits to be obtained in a cost-effective manner, avoiding the need for multiple programmes, reducing transaction costs and programme overlap. Multiple ecosystem service provisions can help ensure that all aspects of an ecosystem on enrolled land are properly managed, increasing the asset value of the ecosystem. PES targeting multiple ecosystem services can enable the landholder to maximise potential payments received, such that conservation becomes more economically feasible, enabling greater ecosystem service provision.

The feasibility of targeting multiple ecosystem services simultaneously depends on the degree of spatial correlation between different types of ecosystem services. Spatial mapping tools help to identify where multiple service benefits coincide. Though there may often be synergies in service provision (*e.g.* avoided deforestation results in both biodiversity and carbon benefits), there are cases when trade-offs can also arise (Nelson *et al.,* 2008). For example, whereas native and mixed crops provide biodiversity benefits, monocultures of fast-growing tree species such as Eucalyptus may provide more rapid carbon sequestration benefits. Farley *et al.* (2005) highlighted this problem in West Africa, where carbon sequestration (*i.e.* afforestation/reforestation) projects can negatively affect water regimes and biodiversity. The ultimate objective of the PES programme must therefore be clear, potential trade-offs recognised, and safeguards may be needed to prevent adverse impacts on other ecosystem services (see OECD/Karousakis, 2009). In this context, environmental benefit indices and scoring approaches become not only a way of evaluating the quality of potential contract benefits, but are mechanisms through which discrete ecosystem service priorities are traded off against each other. Any weights associated with an EBI or scoring mechanism can also be modified in sequential PES sign-up rounds to reconcile trade-offs. This has been done for example in the Mexican PEHS programme (Figure 3.4) where weights have been adjusted over time to better address the policy priorities.

Similar targeting methods have been used to allocate payments in the Socio Bosque programme in Ecuador. Based on a system of scores, land area has been classified into three categories of priority: Priority 1 (scoring between 12.1 to 25); priority 2 (7.1 to 12) and priority 3 (0 to 7). The scores are based on high deforestation pressure, storage of carbon in biomass, water supply and poverty alleviation.

Figure 3.4. Targeting PEHS in Mexico

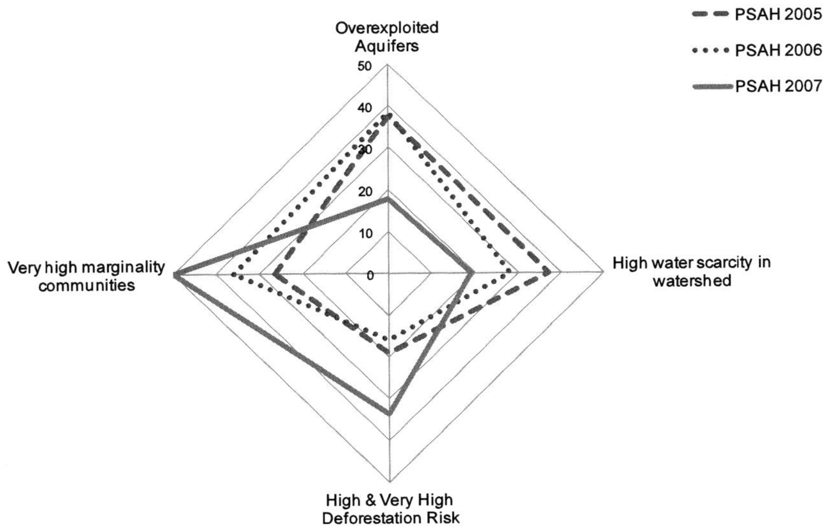

Source: Muñoz Piña *et al.*, 2009.

Though these types of targeting approaches entail higher transaction costs, experience with their use suggests that the resulting cost-effectiveness gains are improved. There are also other types of PES design characteristics that can be introduced in the programme to reduce transaction costs. In the Costa Rican PES for example, private forest landholders are required to have a minimum of one hectare to receive payments for reforestation and two hectares in the case of forest protection. The maximum area for which payments can be received is 300 hectares (and 600 hectares for indigenous peoples' reserves) (Grieg-Gran *et al.,* 2005). Aggregating small projects is also possible to help reduce the transaction costs associated with a payment contract. These types of PES design elements can help to ensure more equitable participation in the PES programme and help to reduce administrative costs.

3.2 Targeting ecosystems services at risk of loss or degradation

In addition to targeting payments to ecosystem services with the highest benefits, it is essential to ensure that any payment leads to additional benefits relative to the business-as-usual (BAU) scenario. For example, payments for habitat protection are only additional if in their absence the habitat would be degraded or lost. Information on the BAU or baseline scenario is critical in ensuring PES additionality. Clear understanding of whether or not ecosystem services are at risk of loss or degradation is therefore needed. Historical and current trend data on biodiversity and ecosystem service loss are a starting point and are needed to develop future reference projections. Though this can be a complex task, there are different ways this can be undertaken. For example, to target PES in Madagascar, Wendland *et al.* (2009) estimate the probability of deforestation (via a multivariate probit model) by examining distance to roads and footpaths, elevation, slope, population density, mean annual per capita expenditure and other characteristics. A similar approach is used to assess deforestation risk in the Mexican PEHS programme. In this case, the variables used to estimate deforestation risk include distance to the nearest town and city, slope, whether it is an agricultural frontier, and if it is located in a natural protected area.

3.3 Targeting providers with low opportunity costs

Finally, PES programmes can increase their cost-effectiveness if, given sites with identical ecosystem service benefits and risk of degradation or loss, payments are differentiated and prioritised to those sites where landholders have lower opportunity costs of alternative land uses. In the Costa Rican PES for example, Wunscher *et al.* (2006) illustrate that differentiating payments according to opportunity costs could allow the enrolment of almost twice the area of land, representing more than double the environmental benefits per cost (Box 3.2).

Obtaining accurate information on ecosystem providers' opportunity costs is not straightforward as they have an incentive to overstate these costs in an effort to extract information rents via higher payments (see Chapter 1). Programme administrators have a number of options to assist revelation of the landholder's true opportunity costs. Specifically, they can gather additional information in the form of costly-to-fake signals or they can use inverse auctions.[6]

Information on ecosystem supplier attributes and activities which are correlated with their opportunity costs can be used to infer the correct price. The information should be based on costly-to-fake signals, for example, distance to markets, current land use, assessed value, or labour and production inputs. Readily available market information can also be used, and incorporated into a model to estimate opportunity costs. In the US Conservation Reserve Programme for example, local land rental rates are combined with information on field soil types, a proxy for productivity, to give a reasonable indication of the opportunity costs of retiring agricultural land. This is then used as a maximum acceptable price, removing the landholders' ability to claim unreasonably high payments. To proxy for opportunity costs in Madagascar, Wendland *et al.* (2009) use data on the opportunity costs of agriculture and livestock produced by Naidoo and Iwamura (2007). Naidoo and Iwamura compiled information on crop productivity and distribution for 42 crop types, livestock density and estimates of meat produced from a carcass, and producer prices to measure the gross economic rents of agricultural land across the globe. Wendland *et al.* clipped this global data to Madagascar's boundaries. Gross economic rents ranged from USD 0 to 529 per hectare for Madagascar, with a mean value of USD 45 per ha, per year. The value of USD 91 per ha, per year (one standard deviation) was used as the cut-off to exclude areas of high opportunity costs.

Box 3.2. **Costa Rica Payments for Environmental Services**

In 1996, Costa Rica replaced an ineffective system of tax deductions for reforestation with a PES programme. Funded by oil tax revenues, the World Bank, the Global Environment Fund, and the German aid agency KfW, the programme enrolls land to protect areas of natural forests, establish sustainable timber plantations, regenerate natural forests, and establish agro-forestry systems. The aim is to incentivise the provision of carbon sequestration, water quality, biodiversity protection, and scenic beauty services on private land.

Between 1997 and 2005 forest protection was supported on 1.1 million acres, and timber plantations on 67 000 acres. The programme gives a uniform per acre payment level irrespective of the quality or quantity of the ecosystem services provided. Contracts are prioritised according to predefined spatial criteria, including, officially acknowledged biological corridors, private property located within protected areas, zones with a low social development index, and expiring contracts (Pagiola, 2006).

Wunscher *et al.* (2006) analysed the Costa Rican PES programme and demonstrated that there are potential gains from employing a more discerning contract selection process, together with differentiated payments. The study focused on the Nicoya Peninsula in the northwest of Costa Rica. Plots were scored, giving equal importance to carbon sequestration, water quality, biodiversity protection, scenic beauty, and poverty alleviation benefits. Three selection processes were simulated for comparison: a baseline scenario designed to match the current system, and two scenarios selecting the highest scoring sites, one with uniform payments, and one with differentiated payments relative to estimated opportunity costs.

Box 3.2 continued over page

Box 3.2. **Costa Rica Payments for Environmental Services** *(cont.)*

The uniform payment scenario enrolled 14% higher benefits than the baseline scenario, at the same cost, while the flexible payment scenario enrolled almost twice the land area (196.8%), giving more than double the benefits (203%). Moreover, the flexible scenario was able to use savings from the efficient pricing of low quality sites to fund the enrolment of higher quality sites.

	Baseline	Uniform Payment	Flexible Payment
Payment	Uniform	Uniform	Differentiated
Selection Criteria	Priority Area	Environmental score	Environmental score
Total Cost (USD)	69 476 (100%)	69 429 (99.9%)	69 471 (99.9%)
Area (ha)	1 736.9 (100%)	1 735.7 (99.9%)	3 417.8 (196.8%)
Environmental Score	27 421 (100%)	31 325 (114%)	55 724 (203%)
Score per USD	0.395 (100%)	0.451 (114%)	0.802 (203%)

However, obtaining information on costly-to-fake signals still incurs research costs. The efficiency of the payment will directly depend on the quality of this research and the strength of the correlation between the signal and the opportunity costs, which must be assessed on a case by case basis.

Exploiting competition between ecosystem service suppliers for conservation contracts through inverse auctions can provide an effective cost-revelation mechanism. Where suppliers are heterogeneous in their opportunity costs, and demand for contracts exceeds supply (*i.e.* the conservation budget), competitive procurement auctions are possible.

The recognition of the potential gains from the use of inverse auctions as a payment allocation mechanism has stimulated heightened interest from policy makers. Though their use in PES programmes is not yet common, they are becoming more widespread in developed and developing countries. Inverse auctions have been used to allocate PES contracts in Australia, Canada, Finland, Germany, Indonesia, Tanzania, the United Kingdom, and the United States (DSE, 2009; Hill *et al,* 2010; Juutinen and Ollikainen, 2010; Latacz-Lohmann and Schilizzi, 2005; Jack, 2009; EAMCEF, 2007; Claassen, 2009). Part II of this book presents three PES case studies that have incorporated inverse auctions in their design.

Notes

1. For additional examples of biodiversity metrics and indicators adopted in the US Conservation Reserve Programme and the Tasmanian Forest Conservation Fund, see Chapters 6 and 7 respectively.
2. http://esd.uvm.edu/
3. http://www.naturalcapitalproject.org/
4. http://rmgsc.cr.usgs.gov/ecosystems/
5. www.ip-sensor.org
6. Screening contracts can also be used in theory, but in practice are complicated; see Ferraro (2008).

References

Claassen R. (2009), USDA Briefing Room – Conservation Policy: Background. http://www.ers.usda.gov/briefing/conservationpolicy/background.htm.

DSE (Deparment of Sustainability and Environment) (2009), ecoMarkets, Virginia, Australia, http://www.dse.vic.gov.au, accessed on 4th April 2010.

EAMCEF (Eastern Arch Mountains Conservation Endowment Fund) (2007), Welcome to the Eastern Arc Homepage, http://www.easternarc.or.tz/, accessed on 4th April 2010.

Farley, K., E. Jobbagy and R. Jackson (2005), "Effects of afforestation on water yield: a global synthesis with implications for forestry", *Global Change Biology*, Vol. 11.

Gorenflo, L.J., K.M Chomitz, C. Corson, G. Harper, M. Honzák, and B. Özler, (forthcoming), "Exploring the Relationship Between People and Deforestation in Madagascar", in R. Cincotta, L.J. Gorenflo, (eds.), *Human Population: The Demography and Geography of Homo sapiens and their Implications for Biodiversity*, Berlin.

Grieg-Gran, M., I. Porras and S. Wunder (2005), "How can market mechanisms for forest environmental services help the poor? Preliminary lessons from Latin America", *World Development*, Vol. 33, No. 9.

Hill M., D. McMaster, T. Harrison, A. Hershmiller, and A. Plews (forthcoming 2010), "A Reverse Auction for Wetland Restoration in the Assiniboine River Watershed", submitted to *Canadian Journal of Agricultural* Economics, Saskatchewan, Canada.

Jack, B.K. (2009), "Auctioning conservation contracts in Indonesia–Participant learning in multiple trial rounds", CID Graduate Student and Research Fellow Working Paper No. 35, Center for International Development at Harvard University, February 2009.

Juutinen A. and M. Ollikainen (2010), "Conservation contracts for forest biodiversity: theory and experience from Finland", *Forest Science*, Vol. 56, No. 2.

Kapos V., C. Ravilious, A. Campbell, B. Dickson, H. Gibbs, M. Hansen, I. Lysenko, L. Miles, J. Price, J.P.W. Scharlemann, and K. Trumper (2008), *Carbon and Biodiversity: a demonstration atlas*, UNEP-WCMC, Cambridge, UK.

Karousakis, K. (2009), *Promoting Biodiversity Co-Benefits in REDD,* OECD Environment Working Paper Series No. 11, OECD, Paris.

Latacz-Lohmann, U. and S. Schilizzi (2005), "Auctions for conservation contracts: a review of the theoretical and empirical literature", Report to the Scottish Executive Environment and Rural Affairs Department.

Naidoo, R. and T. Iwamura (2007), "Global-scale mapping of economic benefits from agricultural lands: implications for conservation priorities", *Biological Conservation*, Vol. 140.

Nelson, E., S. Polasky, D. Lewis, A. Plantinga, E. Lonsdorf, D, White, D. Bael and J. Lawler (2008), "Efficiency of incentives to jointly increase carbon sequestration and species conservation on a landscape", *Proceedings of the National Academy of Sciences*, Vol. 105, No. 28.

Pagiola, S. (2006), "Payments for Environmental Services in Costa Rica", *Munich Personal RePEc Archive.*

Tallis, H.T., T. Ricketts, E. Nelson, D. Ennaanay, S. Wolny, N. Olwero, K. Vigerstol, D. Pennington, G. Mendoza, J. Aukema, J. Foster, J. Forrest, D. Cameron, K. Arkema, E. Lonsdorf and C. Kennedy (2010), InVEST 1.004 beta User's Guide. The Natural Capital Project, Stanford.

Villa, F., M. Ceroni, K. Bagstad, G. Johnson and S. Krivov (2009), "ARIES (artificial intelligence for ecosystem services): A new tool for ecosystem services assessment, planning, and valuation", in Proceedings of the 11th Annual BIOECON Conference on Economic Instruments to Enhance the Conservation and Sustainable Use of Biodiversity.

Wendland, K. (2008), "Rewards for ecosystem services and collective land tensure: lessons from Ecuador and Indonesia", Tenure Brief, Land Tenure Centre, Wisconsin.

Wunder S. and S. Wertz-Kanounnikoff (2009), "Payments for ecosystem services: a new way of conserving biodiversity in forests", *Journal for Sustainable Forestry*, Vol. 28, No. 3.

Wunscher, T., S. Engel and S. Wunder (2006), "Payments for environmental services in Costa Rica: increasing efficiency through spatial differentiation", *Quarterly Journal of International Agriculture*, Vol. 45, No. 4.

Paying for Biodiversity: Enhancing the Cost-Effectiveness of Payments for Ecosystem Services

Chapter 4

Mobilising finance for payments for ecosystem services

This chapter considers the different sources of PES finance, broadly classified as direct user-financing and third-party financing where governments or organisations act on behalf of the beneficiaries. The advantages and disadvantages associated with each are assessed. The motivations for private sector financing of PES programmes are illustrated with examples, highlighting the opportunities and challenges for scaling up private sector engagement.

Identifying sufficient, long-term, and reliable sources of finance is important in order to ensure that the financial resources necessary to carry out the desired environmental objectives can be met in practice. This entails (i) a financial needs assessment; and (ii) a resource mobilisation strategy. This is of particular importance in the context of PES, where continuous payments to landholders may be needed. This chapter considers the different sources of PES finance, broadly classified as user-financed and third-party financed programmes, and the advantages and disadvantages associated with each. It also highlights existing experience and the motivations for private sector financing of PES programmes and considers possible opportunities and challenges for scaling this up.

4.1 Identifying ecosystem service financing needs and sources

Identifying ecosystem service buyers and ensuring sustainable finance for PES is central to the long-term success of the programme. Buyers of ecosystem services can be the users and beneficiaries themselves, or third parties purchasing the service on their behalf. Ensuring sustainable finance for PES is essential – several programmes have been undermined as inadequate attention has been given to this issue. The implementation of a PES programme in Bhopal, India, has failed to come to fruition due to a lack of sustainable finance (Agarwal *et al.*, 2007) while in Ecuador a new financial strategy was required to continue the Pimampiro programme after third party funding ended (Echeverria *et al.*, 2004). Finance for PES is needed to cover different types of costs. These can be classified into two categories: short-term design and capacity building costs; and longer term implementation costs which cover the ecosystem service payments needed to induce the desired behavioural changes in land use decisions.

Financing PES design and capacity building

The PES programme design and capacity building phase may require a relatively large injection of up-front finance. The decision to launch a PES programme will be based upon an existing foundation of research considering the biological patterns and processes, local environmental pressures and the need for the conservation and sustainable use of biodiversity and associated ecosystem services. Additional funds are required to assess the applicability of PES and the optimal design, considering the environmental, economic, and social context. Specifically, up-front costs may include short-term funding for research, stakeholder consultation and the creation of the necessary institutions, including those for legal aspects, contract allocation, and for data collection and monitoring.

The start-up costs for Ecuador's Pimampiro programme were relatively large at USD 38 000, with annual PES payments of about USD 6 000 (Wunder and Ablan, 2008). In contrast, in the Tasmanian FCF the programme transaction costs, including design and capacity building costs, were much lower representing a little over 10% of the AUD 50 million three year budget.

There are a number of programmes launched independently by the private sector. For example, Krakatau Steel, as the service beneficiary, financed research for a watershed management programme in Indonesia, and Nordic Shell Holdings SA, as the service provider, financed the research for their blue mussel farm water purification projects in Sweden (see Table 4.1 for range of examples). Often however, the initial stages of the programme development are undertaken by third parties. In some cases the opportunities provided by PES simply may not have occurred to the potential beneficiaries; as PES programmes continue to proliferate, this effect is likely to diminish. In other cases, the initial research and development costs represent a large financial risk, unacceptable for some individuals and firms (especially those of small to medium size). In Himachal Pradesh, India, for example, the International Institute for Environment and Development (IIED) and Winrock International carried out the necessary research and facilitated negotiations between small scale farmers to secure the implementation of improved upstream watershed management practices benefiting downstream irrigation in the Oach-Kuhan catchment (Agarwal *et al.,* 2007). Without the involvement of these organisations, the transaction costs may have been too great for the individual farmers to set up the programme.

Governments and international organisations also provide finance for the development of PES programmes by supplying some PES programmes with a donation, grant, or loan. The finance for these one-off grants and loans may be sourced from the general budget of governments and international organisations, or from funds ear-marked for conservation and development aid. The Global Environment Facility (GEF) biodiversity mainstreaming portfolio, for example, includes more than 30 projects that apply the PES mechanism. Within these projects the GEF supports the design and implementation of PES schemes to compensate resource managers for off-site ecological benefits. Investments have been made in the development of national systems of PES, regional or local schemes with investments from the private sector, and public-private partnerships (GEF, 2009).

An effective mechanism by which governments can fund conservation and sustainability projects now, but delay the payment until the service is delivered, is through the issue of 'Green' bonds (IFC, 2010). Green bonds respond to the increased demand for environmental investment products, giving private investors a low-risk investment with a fixed return (World Bank, 2010). The bond issuer is typically required to pay the bond investor a fixed-rate annual coupon, plus repay the principle loan at bond maturity. Given their innovative nature green bonds generally offer a higher return than conventional sovereign bonds. Since 2008, the World Bank has issued USD 1.5 billion in AAA/Aaa rated Green Bonds through 20 transactions in 15 different currencies. These have financed projects including watershed management and avoided deforestation PES programmes, as well as other climate change mitigation and adaptation projects (World Bank, 2010).

There is a limit to the capacity of governments and organisations such as the World Bank to continue borrowing to fund conservation and sustainability programmes (including PES start-up costs). To assess the potential of ecosystem service providers to borrow against future PES earnings rather than rely on government funds, EnviroMarket and Forum for the Future (2007) have outlined a proof of concept for Forest-Backed Bonds. These are innovative asset-backed bonds which could be issued directly by sustainable forest managers, or a specialised third party, against a variety of potential cash flows from sustainable forest management. Ecosystem service payments are a potentially important part of sustainable forest management revenues (EnviroMarket and Forum for the Future, 2007; PRP, 2009). In theory this would allow finance to be raised, independently of third party funding, for the development of PES programmes by forest managers expecting to receive ecosystem service payments. Despite the potential, there are a number of practical and theoretical issues that need to be addressed before this can be a reality, as illustrated in Box 4.1.

Box 4.1. **Forest-backed bonds for PES as part of sustainable forest management**

For forest ecosystem service providers to raise capital for sustainable forest management and PES programmes by enticing investment in forest-backed bonds from the private sector, the bonds must be competitive, in terms of return and risk, in comparison to conventional forestry investments, as well as other debt products. Typically sustainable forest management is considered to be low return, high risk, resulting in a poor credit rating and low demand (PRP, 2009), however, there are a number of factors which could lower the risk and increase return.

Box 4.1 continued over page

Box 4.1. Forest-backed bonds for PES as part of sustainable forest management *(cont.)*

The potential rate of return of sustainable forest management is still considerably lower than that associated with conventional forestry (EnviroMarket and Forum for the Future, 2007). For forest-backed bonds to be a success, other sources of cash flow will be required to increase the underlying cash flows of future revenues. For example, from ecosystem service payments, pharmaceutical prospecting concessions, and agro-forestry. Furthermore, the inauguration of a mechanism for avoided deforestation under the UNFCCC carbon negotiations (REDD-plus) would substantially increase the potential return of many sustainable forest management projects.

The potential investment risks associated to forest-backed bonds include political risks in the country of operation, insecure property rights, property loss from human or natural events, market risk from changing product prices, and operational risk from poor management, as well as low investment liquidity. A variety of risk management and mitigation measures were identified including portfolio diversification, insurance, and securitisation (EnviroMarket and Forum for the Future, 2007). Furthermore, while the concept gains acceptability in the investment arena, governments or respected institutions could guarantee the bonds (PRP, 2009). Investors in these 'wrapped bonds' have the assurance of the guarantor that they will cover any losses in case of default by the bond issuer, thus dramatically reducing risk. Government or institutional liability in these products is lower than that from fixed-income bonds issued by them directly because they only have to make a payment if the underlying asset defaults.

If these increases in cash flow and reductions in risk can be achieved, it is likely that in the future there may be considerable demand for forest-backed bonds.

Financing PES programme implementation

PES programme implementation requires a sustainable long-term source of financing to cover the ecosystem services themselves (consisting of the landholders opportunity costs, transaction costs and any management or protection costs), and the programme maintenance costs, including monitoring, reporting, verification and review. PES implementation can be financed by users or beneficiaries, and by third parties acting on behalf of the beneficiaries. Both approaches have been successfully utilised for securing different types of ecosystem services, though there are advantages and disadvantages to each.

Programmes which are financed directly by the users or beneficiaries of the ecosystem service are generally in a better position to negotiate an efficient price because they have direct access to information on the quality of the service provided (Engel *et al.*, 2008; Blackman and Woodward, 2010). Direct beneficiary financing also dispels some of the concerns over finance sustainability because as long as ecosystem service benefits are supplied by the programme, the beneficiaries have an incentive to continue providing finance.

Governments and international organisations have been instrumental in the recent development and proliferation of PES programmes. Such assistance from international organisations is particularly useful for countries with little experience with PES or other market-based mechanisms. In contrast to beneficiary financing, for programmes financed by third parties, the buyers – often governments or institutions – are detached from the service and may not be able to value the service benefits or the magnitude of the demand as accurately. Furthermore, governments may also be influenced by political pressures, and institutions by their financers or shareholders, and their objectives may differ from those of the ecosystem service beneficiaries (Blackman and Woodward, 2010). However, there are also advantages with government-financed PES programmes. In particular, they are likely to benefit from economies of scale. This is because PES programmes can entail large transaction costs, including identifying and matching service providers and users, negotiating conditional contracts, monitoring compliance and enforcing contract terms (Engel *et al.*, 2008; Blackman and Woodward, 2010). Government-financed programmes are able to spread these costs over a large number of agents.

Third-party finance is typically thought to be less sustainable than beneficiary finance. This is because they are susceptible to changes in government administration or funding priorities of organisations (Engel *et al.*, 2008; Blackman and Woodward, 2010). Ideally, programmes would not depend on donations and grants from third-parties beyond the design and capacity building stage, and should instead seek to secure a sustainable source of finance for their continued ability to make ecosystem service payments.

Governments and institutions secure funding support for PES programmes in a number of ways, which can affect how sustainable it is. A budgetary allocation for a programme is often used to secure a nationally relevant service to provide benefits to the wider population. However, such finance can have poor sustainability, especially if there is a risk of government changes or policy reforms (Blackman and Woodward, 2010). Enacting the funding provision in laws or constitutional documents can reduce this risk. In the United States, the Conservation Reserve Programme is allocated funding via the Farm Bill, revised every four to six years (Claassen *et al.*, 2008). Trust funds, with legally binding principles of use, can provide interest payments to increase the sustainability of third-party financing. For example, in Ecuador, a dedicated fund was set up to help sustain the Pimampiro programme after financial support from the Inter-American Foundation ended (Bann, 2003). The capitalised interest on the initial donation is used to finance continued payments, along with a 20% water consumption surcharge for local residents, who benefited from the improved local water services (Box 4.2) (Echeverría *et al.*, 2004; Wunder and Alban, 2008).

Another mechanism used by governments to provide sustainable finance is earmarked taxes or charges. For example, when Costa Rica replaced their forest credit system with the PES programme, it revised the funding from budget allocation to a system largely financed through a 3.5% fuel consumption tax (Wunscher *et al.,* 2006). While the fuel tax is not directly levied on the beneficiaries of the programme, it represents a sustainable source of finance from a related environmentally damaging activity. Moreover, such a tax effectively leverages finance from both the private sector and the public. User charges are often used in watershed-based PES programmes because service consumption is directly measurable; the Mexican Payments for Environmental Hydrological Services (PEHS) is wholly financed by water use charges, with almost 2.5% of annual water revenues earmarked for the PEHS programme (Muñoz Piña *et al.*, 2008). It is important that the conditions of revenue use from taxes or charges are clearly defined and enforced. In Brazil, 5% of the value added sales tax is allocated to municipalities that commit to watershed forest conservation for clean drinking water (May *et al.*, 2002). However, Mayrand and Paquin (2004) note that while the programme is largely successful, some municipalities have used the funds for non-conservation objectives.

The geographical scale of the ecosystem services benefits has implications for the appropriate scale of PES finance. Ecosystem service benefits are provided locally, nationally and internationally (Figure 4.1). Mobilising user finance therefore depends on the geographical scale of the ecosystem service benefits that are being provided. To create the most direct link between the service providers and beneficiaries, the geographical scale of the financing should match that of the service provision. For example, if the objective is to address the local public good benefit of watershed services, the most appropriate finance may be that from beneficiaries within the watershed; if the objective is to address a nationally or internationally relevant service, it may be more appropriate to mobilise PES finance at the national and international level, respectively.

Box 4.2. Creating funds to finance PES in Ecuador and Tanzania

The Ecuadorian Pimampiro payments for watershed service programme has successfully capitalised an initial donation of USD 15 000 from the Inter-American Foundation (IAF) and the UN Food and Agriculture Organization (FAO). Investing in a simple savings account gives annual returns of 4 to 10% and after five years the fund grew to nearly USD 20 000. Together with a 20% water consumption surcharge on 1 350 households in Pimampiro, the fund helps sustain the programme's ability to continue ecosystem service payments to the Nueva América community for service provision, as illustrated in the following diagram.

Organisation of the Pimampiro fund

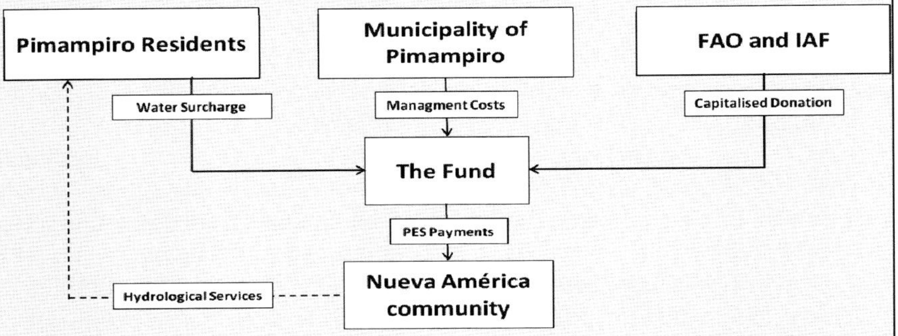

Ideally funds should be created with strict principles of use, through a trust fund, to ensure that the money is not diverted to other ends. Despite the financial success of the fund, Wunder and Alban (2008) note that a lack of such principles could potentially threaten the sustainability of the fund.

A trust fund has been used in the Tanzanian Eastern Arc Mountains to deliver finance for a number of long-term conservation and forest biodiversity management programmes within the region. The Trust Fund has strict guiding principles and funding eligibility criteria, ensuring the finance is only directed to conservation projects that meet these criteria. It is a joint initiative of the Government of the United Republic of Tanzania, the World Bank and the GEF.

The programme received a USD 7 million grant from the GEF with which it set up the investment Trust Fund in 2006. As the interest from the fund alone was not considered sufficient to meet the goals of the programme, it was decided to invest the fund in the capital markets through a leading investment bank with an aim of achieving higher growth. By the end of June 2008 the funds had grown to USD 7 303 020. However the investments were hit by the global economic recession and slumped to USD 5 849 398 by the end of the year. The investment recovered to USD 6 540 250 by the end of June 2009 (EAMCEF, 2007).

Trust Funds with strict guiding principles of use are an effective way of using grants and donations to fund PES programmes in the longer term. However, the risk exposure associated with the type of investment, from low risk savings accounts, to more risky investments in financial markets, need to be carefully considered.

Figure 4.1. **Stylistic representation of the spatial scale of different ecosystem service benefits**

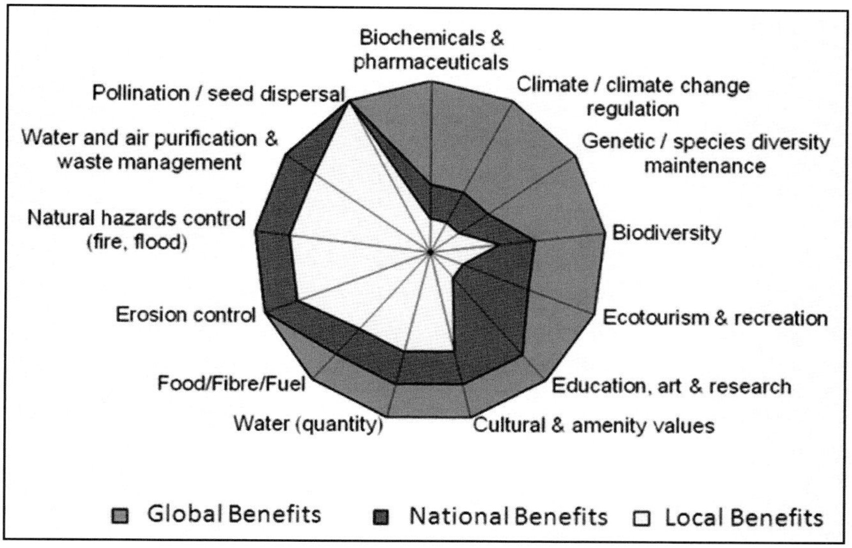

Source: TEEB, 2009.

In some cases however, it may not be practical or cost effective to obtain finance at the corresponding geographical scale. For example, it was suggested that the Mexican PEHS should allocate the funds to the regional watershed programmes in the same geographical proportions as the federal water fees were collected from watersheds. However, this was not carried out because the majority of fees were collected from a small number of urban areas, which were not necessarily located in the watersheds in greatest need of ecosystem service payments (Muñoz Piña *et al.*, 2008).

4.2 Experience with private sector PES financing

The need to better engage and leverage finance from the private sector in biodiversity conservation and sustainable use is being increasingly recognised (CBD, 2010; UNEP, 2008). In the context of PES, there are a growing number of programmes that are financed voluntarily by private firms and individuals (see Table 4.1). These are often smaller scale programmes providing localised ecosystem service benefits to firms nearby. These programmes resemble Coasian bargaining, conforming most closely to the PES definition presented in Chapter 1.

Table 4.1. **Finance sources across a selection of PES programmes**

PES Example	Ecosystem Service Provision and Financing Sources	Scale of Service	Scale of Finance
	PRIVATE FIRMS AND INDIVIDUALS		
Tmatboey, Cambodia (Clements *et al.*, 2008)	**Biodiversity conservation**: Tourists pay to view key avian species (USD 30 if all viewed, USD 15 if subset viewed), financing conservation and land management plans by villages.	Local and international	International
PROFAFOR, Ecuador (Wunder and Alban, 2008)	**Carbon sequestration**: FACE, a consortium of Dutch electricity companies, pays for afforestation and reforestation to offset emissions from a new power plant (80% of sequestration is not eligible under Kyoto because contracts signed before Kyoto came into force).	International	International
Sierra de las Minas Reserve, Guatemala (IIED, 2007)	**Hydrological services**: Downstream water users (Coca Cola, PAINSA paper plant, Licorera Zacapaneca Rum distillery, and hydroelectric plants) pay for watershed management on upstream land to ensure flow of useable water. Design and capacity building costs from WWF, UNEP, and other international donors.	Local	Local
Vosges Mountains, France (Perrot-Maitre, 2006)	**Hydrological services**: Nestle-Vittel pay to change management practices on farms, reducing contamination of the aquifer and increasing water quality.	Local	Local
Gavot plateau, France (WWF, 2006)	**Hydrological services**: Danone-Evian pays (two thirds) farmers to reduce fertiliser use, reducing contamination of the waterways and lowering purification costs. **Environmental quality**: Local residents pay (one third) farmers to implement management practices to maintain the environmental quality of the area.	Local	Local
Himachal Pradesh, India (Agarwal *et al.*, 2007)	**Hydrological services**: downstream famers in the Oach-Kuhan catchment finance watershed management on upstream farms to reduce sedimentation of irrigation reservoir, via water use charges. Design and capacity building costs from IIED and Winrock International.	Local	Local

Table 4.1 continued over page

Table 4.1. **Finance sources across a selection of PES programmes**
(cont.)

PES Example	Ecosystem Service Provision and Financing Sources	Scale of Service	Scale of Finance
	PRIVATE FIRMS AND INDIVIDUALS		
Cidanau, Indonesia (Munawir and Vermeulen, 2007)	**Hydrological services**: Krakatau Steel, which has a long history of working with local governments on conservation and hydrological research programmes, pays for watershed management on upstream farms to reduce reservoir sedimentation. Facilitation costs supported by IIED and LP3ES.	Local	Local
Sasumua, Kenya (Planning stage) (Mwengi, 2008)	**Hydrological services**: Private water plant will pay for watershed management on upstream farms to reduce processing costs. Design and capacity building costs from World Bank and ICRAF.	Local	Local
Lake Naivasha Watershed, Kenya (Mwengi, 2008)	**Hydrological services**: Downstream water users (sewage company, commercial flower growers, and geothermal electricity plant) pay for watershed management on upstream land to ensure flow of useable water. **Environmental quality**: Tourist industry pays for watershed management to reduce pollution and conserve the lake environment. Design and capacity building costs from WWF and CARE.	Local	Local
Panama, Canal (UNEP, 2008)	**Hydrological services**: A reinsurance firm issued a bond, the revenues from which support watershed reforestation by local communities, reducing liability risks associated with canal dredging and closure. Canal users buy the bond in return for reduced insurance premiums.	Local	Local

Table 4.1 continued over page

Table 4.1. **Finance sources across a selection of PES programmes** *(cont.)*

PES Example	Ecosystem Service Provision and Financing Sources	Scale of Service	Scale of Finance
	PRIVATE WITH GOVERNMENT / NGO / INTERNATIONAL DONOR		
PES, Costa Rica (Wunscher *et al.*, 2006; Pagiola, 2006)	**Biodiversity conservation / Carbon sequestration / Environmental quality / Hydrological services**: A 3.5% fuel consumption tax, principally, as well as a water use tariff, credits from the CDM market, voluntary contributions from hydroelectric companies and agri-businesses, and donations from international organisation and governments, pay for avoided deforestation on private land, securing a range of benefits. Design and capacity building costs from the World Bank, GEF, and German KfW Bank.	National, local and international	Principally national and local, with some international
Pimampiro, Ecuador (Wunder and Alban, 2008)	**Hydrological services**: Interest on capitalised donation from FAO and Inter-American Foundation, as well as increased water use charges pay the Nueva America Autonomous Association, made up of numerous individual landholders, some of which receive payments, for avoided deforestation and the implementation of conservation land management practices. Design and capacity building costs from FAO and Inter-American Foundation.	Local	Local and international
PEHS, Mexico (Muñoz Piña *et al.*, 2008)	**Hydrological services**: Charges from federal water use, and additional funding from the government budget, pay for avoided deforestation and the implementation of watershed management practices to ensure the provision of clean water downstream.	Local	National
Danube, Romania and Bulgaria (GEF, 2009)	**Biodiversity conservation / Environmental quality / Hydrological services**: GEF (42%) and WWF (47%), the national and local governments, fish farms and the tourism sector (11%), pay for watershed management practices to conserve the environmental quality of the area. Design and capacity building costs financed by the WWF.	Local and international	Principally international, with some local and national

Table 4.1 continued over page

Table 4.1. **Finance sources across a selection of PES programmes**
(cont.)

PES Example	Ecosystem Service Provision and Financing Sources	Scale of Service	Scale of Finance
	GOVERNMENT		
ecoMarkets, Australia (EcoTender, BushTender, BushBroker) (DSE, 2009)	**Biodiversity conservation / Carbon sequestration / Environmental quality / Hydrological services**: Victoria State Government pays private landholders to conserve and enhance the environment on their land.	Local and national	National
OPUL, Austria (Lebensministerium, 2010)	**Agri-environmental quality**: the national government pays farmers for improved environmental stewardship on their land, in particular a reduction in agricultural intensity and the maintenance of natural resources. Receives funding from the EU Common Agricultural Policy.	National	National and international
Slopping land conservation programme, China (Bennett, 2008)	**Erosion reduction**: the National Government pays private land owners / occupiers to plant trees on slopping land to reduce erosion and land degradation, with an emphasis on income support.	Local and national	National
Amfissa, Greece (Vakrou, 2010)	**Agri-environmental quality**: the National Government pays farmers for maintaining a region of unique 150 year old olive groves. Receives funding from the EU Common Agricultural Policy.	Local and international	National and international
Nordic Shell Holdings, Sweden (Zandersen *et al.*, 2009)	**Hydrological services**: The municipal water treatment plant pays Blue Mussel producer for effluent purification, avoiding traditional technology expenditures, and improving the water quality in an important marine nature reserve. Design and capacity building costs supported by Nordic Shell Holdings.	Local	Local
Ecological Compensation Areas, Switzerland (SFSO, 2007)	**Agri-environmental quality**: the National Government pays farmers for improved environmental stewardship on their land, in particular a reduction in agricultural intensity and the maintenance of natural resources.	Local and national	National

Table 4.1 continued over page

Table 4.1. **Finance sources across a selection of PES programmes**

(cont.)

PES Example	Ecosystem Service Provision and Financing Sources	Scale of Service	Scale of Finance
	GOVERNMENT		
Rural Development Programme for England, UK (Defra, 2009)	**Agri-environmental quality:** the National Government pays farmers for improved environmental stewardship on their land, in particular the maintenance of traditional agricultural landscapes and natural resources. Receives funding from the EU Common Agricultural Policy.	National	National and international
CRP, US (Claassen *et al.*, 2008)	**Agri-environmental quality / Biodiversity conservation / Carbon sequestration / Hydrological services**: the National Government pays farmers for retiring their land, and implementing environmental management practices.	Local and national	National
Kanagawa, Japan (Hayashi and Nishimiya, 2010)	**Hydrological services / Environmental quality**: Ear-marked income tax and water consumption tax funds Five Year Action Plan for watershed and forest management.	Local	Local
Canada, Assiniboine Watershed (Hill *et al*, 2010)	**Hydrological services / Biodiversity conservation**: the National Government pays landholders to restore the quality of the wetlands providing conservation benefits. Programme in association with NGO's Ducks Unlimited and the Assiniboine Watershed Stewardship Association.	Local and national	National
Tir Gofal, Wales (Welsh Assembly Government, 2007)	**Agri-environmental quality**: the National Government pays farmers for improved environmental stewardship on their land, in particular the maintenance of traditional agricultural landscapes and natural resources. Receives funding from the EU Common Agricultural Policy.	National	National and international

Table 4.1 continued over page

Table 4.1. **Finance sources across a selection of PES programmes**
(cont.)

PES Example	Ecosystem Service Provision and Financing Sources	Scale of Service	Scale of Finance
	GOVERNMENT WITH NGO / INTERNATIONAL DONOR		
Dominican Republic, Upper Sabana Yegua (Gutman and Davidson, 2007)	**Biodiversity conservation / Carbon sequestration / Environmental quality / Hydrological services:** the National Government (71%), with support from GEF (14%), Sur Futuro (11%), and Kellogg Foundation (5%) pay private landholders and farmers to change their land use practices, reducing land degradation and providing a range of environmental benefits, whilst securing natural capital to promote income generation and the provision of basic services.	Local, national and international	National and international
Arabuko Sokoke Forest, Kenya (Mwengi, 2008)	**Biodiversity conservation / Environmental quality**: the National Government, with support from USAID, Birdlife International, and the WWF pay private landholders for forest conservation, afforestation and the implementation of biodiversity enhancing management practices.	Local, national and international	National and international
Eastern Arc Mountains, Tanzania (EAMCEF, 2007)	**Hydrological services**: the National Government set up the Conservation Endowment Fund, with support from a World Bank loan, and a GEF grant, and operational assistance from the UNDP and the IUCN, to pay private landholders to implement improved management practices in the Ruvu River Basin (as well as other projects).	Local, national and international	Local, national and international

Source: OECD, 2010.

Voluntary private sector participation in PES programmes is motivated by several factors, including cost savings, value added to output, improved public relations, and the ability to influence potential future regulations (Gutman and Davidson, 2007). Ensuring the provision of ecosystem services can result in considerable *cost savings* to production processes. Water quality service programmes are particularly well advanced in this area because water is a particularly tangible ecosystem service and an important production input. For example, hydroelectric companies finance sustainable forest management in Kenya and Costa Rica to reduce erosion and avoid the costs of reservoir dredging (Mwengi, 2008; Wunscher *et al.,* 2006). Drink producers such as Nestle-Vittel and Danone-Evian in France, and Coca Cola and Zacapaneca Rum in Guatemala, save water purification costs through improved upstream watershed management (Perrot-Maitre, 2006; WWF, 2006; IIED, 2007). In Lysekil fjord, Sweden, the local waste water plant saves EUR 100 000 per year in traditional technology costs by paying Nordic Shell Holdings SA for water filtration services provided by its Blue Mussel farms. Nordic Shells business plan is based on its ability to produce high quality shellfish while simultaneously delivering ecosystem services (Zandersen *et al.,* 2009).

Insurance companies have also been motivated by cost savings to participate in PES. Many ecosystem services provide buffers against natural hazards, or maintain the economic viability of operations. For example, the loss of wetlands around the Louisiana coast exacerbated the damage caused by Hurricane Katrina (US EPA, 2006). In Panama a reinsurance firm, ForestRE, has established a watershed protection programme to reduce its liabilities from dredging costs and the risk of canal closure (UNEP, 2008).

Firms can also secure *value added to output* goods and services by participating in PES programmes. Organic and certified markets, such as forestry, are growing at 10% a year (Gutman and Davidson, 2007), with consumers increasingly aware of the environmental impacts of their purchases. Agri-environmental PES programmes support the transition from intensive agriculture to organic production throughout Europe. Furthermore, Wunder (2006) notes that certified products produced under Sustainable Forest Management programmes are a form of PES, where the consumer selects certified products, voluntarily paying a premium for the conservation benefits of the sustainable production practices. Veisten (2007) estimated the extra median willingness to pay for eco-labelled IKEA wooden furniture, finding consumers are willing to pay an additional 16% compared to the price of existing unlabelled alternatives. Tourism is another growth sector which is benefiting from PES programmes. For example, hotels are contributing to the funding for a PES programme operating in the Romanian and Bulgarian sections of the Danube to conserve the environmental quality of the watershed (GEF, 2009).

Private sector financing of PES programmes can be motivated by *public relations* concerns and an ambition to improve a firm's image or ensure social acceptability in the region of operation. In the Costa Rican PES, for example, where more than 40 different firms have made contributions totalling over USD 8 million to date, Blackman and Woodward (2010) find that this is motivated by a will to provide "forest protection and provision of environmental services", but also to improve relations with local communities and governments. Payments from tourists to villages in Cambodia, subject to wildlife viewing, not only incentivise environmental protection but also serve to increase the locals' acceptance of tourists' visits to their villages.

Private sector financing may also be motivated by the desire to *delay or influence any future regulation.* Early action can give a strategic advantage by allowing firms to delay or negotiate the final form of subsequent regulations, and also through a first mover advantage (Maxwell *et al.*, 1998). Companies that fail to track current regulation and predict future developments risk competitive disadvantage (Esty and Winston, 2006).

There is considerable scope for scaling-up private sector financing in PES programmes, especially as business becomes more aware of the opportunities that investment in ecosystem services can offer. It is reasonable to expect that most voluntary private sector engagement in PES will focus on opportunities where they can reap the benefits directly, such as through local watershed PES schemes and the sale of organic products. However, voluntary private sector finance in programmes addressing ecosystem service benefits at regional and global scale, such as biodiversity, is still insufficient to address the level of the market failure. Ecosystem service benefits accruing at larger geographic scales are subject to greater free-riding[1] incentives, particularly for ecosystem services that provide non-use values. Thus, leveraging finance for PES via fees and taxes, such as in the Costa Rican and Mexican programmes, is perhaps a more effective way of mobilising finance, including from the private sector.

Notes

1. Free-riding is associated with the public good nature of biodiversity. Individuals or firms have low incentives to pay for the provision of biodiversity and ecosystem service because others cannot be excluded

from enjoying the benefits. This leads an under supply of ecosystem services.

References

Agarwal, C., S. Tiwari, M. Borgoyary, A. Acharya, and E. Morrision (2007), *Developing markets for watershed services and improved livelihoods. Fair deals for watershed services in India,* Natural Resource Issues, No. 10, International Institute for Environment and Development, (IIED), London, United Kingdom.

Bann, C. (2003), Ecuador – "Pimampiro payments for watershed services scheme", in Gutman, P. (ed.), *From Goodwill to Payments for Environmental Services*, WWF.

Bennett, M. (2008), "China's sloping land conversion program: Institutional innovation or business as usual?", *Ecological Economics*, Vol. 65.

Blackman, A, and R. Woodward (2010). "User Financing in a National Payments for Environmental Services Program: Costa Rican Hydropower", Resources for the Future. Washington DC.

CBD (2010), Biodiversity and Business, UNEP/CBD/WG-RI/3/2/Add.2, 25 March 2010.

Claassen R., A. Cattaneo and R. Johansson (2008), "Cost-effective design of agri-environmental payment programs: U.S. experience in theory and practice", *Ecological Economics*, Vol. 65.

Clements, T., J. Ashish, K. Nielsen, D. An, S. Tan, and E. Milner-Gulland (2010), "Payments for biodiversity conservation in the context of weak institutions: Comparison of three programs from Cambodia", *Ecological Economics*, Vol. 69.

Defra (Department for Environment Food and Rural Affairs, UK) (2009), ERDP Schemes, http://www.defra.gov.uk/rural/rdpe/erdp/schemes/index.htm, accessed on 4th April 2010.

EAMCEF (Eastern Arch Mountains Conservation Endowment Fund) (2007), Homepage, http://www.easternarc.or.tz/, accessed on 4th April 2010.

Echeverria, M., L. Vogel, M. Alban, and F. Meneses (2004), *The impacts of payments for watershed services in EcuadorEmerging lessons from Pimampiro and Cuenca*, International Institute for Environment and Development (IIED), London, United Kingdom.

Engel, S., S. Pagiola and S. Wunder (2008), "Designing payments for environmental services in theory and practice: An overview of the issues", *Ecological Economics*, Vol. 65.

EnviroMarkets and Forum for the Future (2007), Forest-backed bonds proof of concept, IFC, London.

Esty, D. and A. Winston (2006), *Green to Gold, How smart companies use environmental strategy to innovate, create value, and build competitive advantage*, Yale University Press, New Haven.

GEF (Global Environment Fund) (2009), Request for CEO Endorsement/Approval. GEF Washington DC.

Gutman P. and S. Davidson (2007), *The Global Environmental Facility and payments for ecosystem services, a review of current initiatives and recommendations for future PES support by GEF and FAO programs,* FAO and WWF.

Hayashi, K. and H., Nishimiya (2010), *Good practices of payments for ecosystem services in Japan – DRAFT*, submitted to TEEB-D2, 7th March 2010, Nagoya, Japan.

Hill M., D. McMaster, T. Harrison, A. Hershmiller and A. Plews (forthcoming 2010), "A Reverse Auction for Wetland Restoration in the Assiniboine River Watershed", submitted to *Canadian Journal of Agricultural* Economics, Saskatchewan, Canada.

IIED (2007), Guatemala – Sierra de las Minas. http://www.watershedmarkets.org/casestudies/Guatemala_Sierra_Minas.html, accessed on 4th April 2010.

IFC (International Finance Corporation) (2010), IFC Green Bonds, Press Releases, www.ifc.org, accessed on 16th April 2010.

Levensministerium (2010), Austria Agri-Environmental Programme (OPUL), http://land.lebensministerium.at/article/articleview/37214/1/5849/, accessed on 4th April 2010.

May, P., F. Neto, V. Denardin, and W. Loureiro (2002), "Using fiscal instruments to encourage conservation: municipal responses to the 'ecological' value-added tax in Parana and Minas Gerais, Brazil" in S. Pagiola, J. Bishop, and N. Landel-Mills (ed.), *Selling forest environmental services: market-based mechanisms for conservation and development*, Earthscan Publications Ltd.

Mayrand K. and M. Paquin (2004), *Payments for environmental services: a survey and assessment of current schemes,* Commission for Environmental Cooperation of North America.

Maxwell J., W., Lyon, and S. Hackett (1998), "Self-Regulation and Social Welfare: The Political Economy of Corporate Environmentalism," Nota di Lavoro 55.98.

Munawir, and S. Vermeulen (2007), *Developing markets for watershed services and improved livelihoods. Fair deals for watershed services in Indonesia,* IIED, London.

Muñoz Piña, C., A. Guevara, J. Torres and J. Brana (2008), "Paying for the hydrological services of Mexico's forests: analysis, negotiations and results", *Ecological Economics*, Vol. 65.

Mwengi, S. (2008), *Payments for ecosystem servies in East and Southern Africa : Assessing prospects and pathways forward,* Katoomba Group.

Pagiola, S. (2006), "Payments for Environmental Services in Costa Rica", *Munich Personal RePEc Archive.*

Perrot-Maitre, D. (2006), "The Vittel payments for ecosystem services: a "perfect" PES case?", IIED, DFID, London.

PRP (Prince's Rainforests Project) (2009), "An Emergency Package for Tropical Forests", PRP, London.

SFSO (2007), *Ecological Compensation Areas, basic data from biodiversity monitoring BDM,* Swiss Federal Statistics Office, Federal Office for the Environment, Switzerland.

TEEB (2009), The *Economics of Ecosystem and Biodivesity for Policy Makers*, TEEB, Bonn.

UNEP (United Nations Environmental Programme) (2008), *Payments for Ecosystem Services, Getting Started, A Primer*, Washington DC.

US EPA (US Environmental Protection Agency) (2006), *Wetlands: Protection property and life from flooding*, Washington DC.

Vakrou, A (2010), "Payments for ecosystem services (PES): Experiences in the EU", presentation given during OECD WGEAB Workshop, 25th March 2010, Paris, www.oecd.org/env/biodiversity.

Veisten, K (2007), "Willingness to pay for eco-labelled wood furniture: Choice-based conjoint analysis versus open-ended contingent valuation", *Journal of Forest Economics*, Vol. 13, No. 1.

Welsh Assembly Government (2007), Tir Gofal, http://wales.gov.uk/topics/environmentcountryside/farmingandcountryside/farming/agrienvironmentschemes/tirgofal/?lang=en, accessed on 4th April 2010.

World Bank (2010), World Bank Green Bonds, http://treasury.worldbank.org/cmd/htm/WorldBankGreenBonds.html, accessed 10th January 2010.

Wunder, S. and M. Albán (2008), "Decentralized payments for environmental services: the cases of Pimampiro and PROFAFOR in Ecuador", *Ecological Economics*, Vol. 65, No. 4.

Wunscher, T., S. Engel and S. Wunder (2006), "Payments for environmental services in Costa Rica: increasing efficiency through spatial differentiation", *Quarterly Journal of International Agriculture*, Vol. 45, No. 4.

WWF (2006), "Private sector – NGO forum to promote ecosystem services and payments for ecosystem servies", presentation by Danone-Evian, http://www.panda.org/what_we_do/how_we_work/policy/development_poverty/macro_economics/?uNewsID=88060, accessed on 4th April 2010.

Zandersen, M., K. Braten, H. Lindhjem (2009), "Payment for and Management of Ecosystem Services, issues and options in the Nordic context", Nordic Council of Ministers, Copenhagen.

Paying for Biodiversity: Enhancing the Cost-Effectiveness
of Payments for Ecosystem Services

Chapter 5

Insights for international payments for ecosystem services

This chapter considers how the insights provided by local and national PES programmes apply to international payments for ecosystem services. IPES refer to programmes where the buyers and sellers of ecosystem services cross jurisdictional boundaries. The chapter discusses IPES-like programmes that are emerging for carbon-related ecosystem services and how international payments for biodiversity and other non carbon-related ecosystem services can be designed and implemented.

Many of the criteria and insights derived for designing and implementing effective local and national PES programmes are also relevant to international PES (IPES). This chapter highlights considerations that are particular to IPES programmes. It discusses recent IPES initiatives in the context of climate change, how these can be designed to promote biodiversity co-benefits, and some of the insights that could be applied to IPES that target biodiversity specifically.

IPES apply the same concept to direct transfers between buyers and sellers of ecosystem services at the international level. A key distinction between PES and IPES is in the types of ecosystem services that each is most suited to target. Ecosystem services occur at different spatial scales, and these scales can be reflected in the design of instruments intended to capture these services. Domestic PES programmes typically focus on services that generate benefits at local or regional levels, such as hydrological regulation, erosion prevention, and aesthetic improvements (*i.e.* landscape beauty) (see also Figure 4.1). In contrast, international financiers are well-positioned to focus on services such as carbon sequestration, genetic information, and non-use values that national government and domestic private sector stakeholders have less incentive to finance due to their global public good characteristics (Klemick and Simpson, 2010).

Examples of existing IPES-like activities include afforestation and reforestation projects under the Clean Development Mechanism (CDM), and more broadly, bio-prospecting arrangements. These mechanisms have also been successful in leveraging finance from the private sector, albeit for different reasons. In the case of the CDM, the private sector is motivated by lower cost greenhouse gas (GHG) emission reductions. These are offset against the mandatory emission reduction targets which many developed countries have agreed to under the Kyoto Protocol of the United Nations Framework Convention on Climate Change (UNFCCC). In the case of bio-prospecting, the private sector is motivated by the value-added that genetic information provides for pharmaceutical and bio-engineering purposes.

5.1 Harnessing synergies between global carbon finance and biodiversity

A new mechanism, Reducing Emissions from Deforestation and forest Degradation (REDD-plus) in developing countries is being proposed under the UNFCCC to help address the global climate change challenge. Successful agreement on a future REDD-plus mechanism would represent a substantial and unprecedented development in the creation of an international financing mechanism to help internalise the carbon-related ecosystem services from forests.

A REDD-plus mechanism is also likely to create substantial co-benefits for other, non-carbon ecosystem services, that forest provide, including biodiversity. Moreover, biodiversity co-benefits can be enhanced if REDD-plus finance is targeted to areas where both high carbon and high biodiversity benefits overlap in space. This would channel REDD-plus finance so that two global ecosystem service benefits could be achieved at the price of one.

In addition to enhancing the biodiversity co-benefits that could be harnessed via a REDD-plus mechanism, supplemental co-financing from biodiversity investors (via bundling or layering) could enable biodiversity benefits to be targeted directly (Karousakis, 2009). Voluntary initiatives to bundle carbon and biodiversity benefits in REDD-plus are already emerging. Examples include the Climate, Community and Biodiversity Alliance (CCBA), which has established standards and criteria to meet these multiple objectives. So-called green REDD-plus credits, entailing premiums for the additional biodiversity benefits they provide, are being purchased on the voluntary carbon market. Such voluntary initiatives to capture the global public good benefits of biodiversity are important – as experience with them grows, they can provide lessons for how they can be improved. Such voluntary biodiversity schemes are unlikely however, to provide the scale necessary to create global demand for biodiversity and change land prices fundamentally (Blom *et al.*, 2008). Just as demand for carbon allowances, CDM credits, and potentially REDD-plus credits in the future, are driven by legally-binding GHG emission reduction commitments and regulated via an international carbon market, large scale international demand for biodiversity conservation and sustainable use would stem from large scale regulatory policies.

5.2 International payments for biodiversity

In this context, recent proposals for an IPES mechanism for biodiversity include a Green Development Mechanism (GDM). The GDM highlights the need to engage and leverage finance from the private sector, and proposes to establish a standard and accrediting process to certify the supply of biodiversity-protected areas. According to the proposal, verification could be undertaken by an independent third party review. By facilitating a functional market, a GDM would enable the sale of certified biodiversity conservation to willing buyers, including businesses and individuals. The proposal suggests to begin with a voluntary phase to pilot the mechanism. This would therefore be analogous to the REDD demonstration activities that are underway to pilot GHG emission reduction activities in the context of avoided deforestation.

Another element of REDD-plus that may be relevant in the context a GDM for biodiversity is the financing approach that is being proposed under the UNFCCC for REDD-plus. Recognising the challenges associated with monitoring emission reductions from deforestation and degradation in developing countries, REDD-plus finance is proposed to be delivered in a three phased approach: (i) for capacity-building (*e.g.* to establish a REDD-plus baseline and monitoring) and the development of a national REDD-plus strategy; (ii) for proxy-based payments (*e.g.* based on area of avoided deforestation); and (iii) for verified emission reductions.

In many countries, the challenges associated with monitoring biodiversity loss and degradation are at least as great, if not greater, than those for monitoring GHG emission reductions from deforestation in developing countries. This is due mainly to the multidimensionality of biodiversity and hence the lack of a single agreed metric or indicator for biodiversity. For a GDM to operate at the international scale, providing certainty to investors on what they are paying for, agreement would be needed on how to quantify a GDM certificate, and thus how to monitor, report and verify (MRV) the biodiversity benefits. A GDM certificate could, for example, provide continuous incentives for improvement by setting up two-levels of compensation, one for proxy-based biodiversity payments - which would be discounted according the uncertainty inherent with the proxy, and a second, higher-level of compensation associated with more rigorous MRV methodologies.

It is important to also note that many local and national PES programmes contribute to the provision of global ecosystem services, concurrently with local services. Such programmes provide international investors the opportunity to co-finance activities as one approach to IPES. One can envision agreements whereby national governments would make concerted efforts to establish well-designed and effective domestic PES programmes (to internalise local and regional external ecosystem benefits), and that these efforts could be layered with international payments to internalise global environmental benefits (such as biodiversity and carbon sequestration) (Karousakis and Corfee-Morlot, 2007). One example of where this has been undertaken is in a recently established PES programme in the Los Negros valley in Bolivia. The programme involves the simultaneous purchase of two ecosystem services, watershed protection and bird habitat. While downstream irrigators through the Municipality of Pamagrande are paying for watershed services, the US Fish and Wildlife Service is paying for the protection of habitat for migratory bird species (Asquith *et al.*, 2008).

A similar approach is being proposed in the Socio Bosque Programme in Ecuador which aims to address deforestation. In addition to the funds allocated to Socio Bosque by the Government of Ecuador, the programme seeks complementary financial stability through a trust fund created within the National Environmental Fund (Fondo Ambiental Nacional, FAN). Through this fund, donations can be received from countries or organisations, as well as economic incentives from a possible REDD-plus mechanism.[1] If, for example, the targeting criteria used in the Socio Bosque programme (which currently prioritises areas with the highest deforestation threat, areas with high carbon storage and other ecosystem services, and areas with the highest levels of poverty) were to also include prioritising areas with high biodiversity benefits, this could open up an additional source of finance, namely from international investors interested specifically in biodiversity conservation and sustainable use.

Finally, it is important to note that the development of any future international mechanisms to help address biodiversity loss and degradation should be supplemented by a more comprehensive system to measure, report and verify existing and new financial flows towards biodiversity. This would help to better identify where the largest financial gaps are, and thus help to target biodiversity finance more effectively.

Notes

1. http://www.ambiente.gov.ec/

References

Asquith N.M, Vargus M.T., Wunder S. (2008). "Selling two environmental services: In-kind payments for bird habitat and watershed protection in Los Negros, Bolivia", *Ecological Economics* 65, No. 4.

Blom, M., G. Bergsma and M. Korteland (2008), *Economic instruments for biodiversity: Setting up a Biodiversity Trading System in Europe,* Delft, CE Delft. Prepared for the VROM.

Karousakis, K. (2009), *Promoting Biodiversity Co-benefits in REDD,* OECD Environment Working Paper Series No. 11, OECD, Paris.

Karousakis, K. and J. Corfee-Morlot (2007), "Financing Mechanisms for REDD: Issues in Design and Implementation", OECD, Paris.

Klemick H. and R.D. Simpson (2010), "International financing for biodiversity conservation: Overview of innovative approaches and persistent challenges", OECD, Paris.

Part II

Payments for ecosystem services programmes case studies

Paying for Biodiversity: Enhancing the Cost-Effectiveness of Payments for Ecosystem Services

Chapter 6

United States: The USDA Conservation Reserve Programme

This chapter presents the design and implementation of the USDA Conservation Reserve Programme, a national agri-environmental programme that provides payments to landholders to retire farmland and improve the environmental quality of agricultural land. The CRP implements a range of management practices to protect highly erodible and environmentally sensitive land, improve water quality, and enhance wildlife habitat. The programme allocates contracts via an auctioning mechanism, targeting payments according to environmental benefits and cost. This helps enhance the cost-effectiveness of the programme. The challenges and lessons learned from the CRP are discussed.

6.1 Introduction

Habitat loss through agriculture is the primary cause of global biodiversity loss (OECD, 2008a; IUCN, 2009a). The United States, where agriculture covers over half the land area,[1] is home to 1192 Threatened[2] species, more than any single nation after Ecuador (IUCN, 2009b). The Conservation Reserve Program (CRP) is one of the main mechanisms through which biodiversity loss from agriculture is combated in the United States; its stated goals are to protect highly erodible and environmentally sensitive cropland.

The CRP, initiated in 1985, is primarily a land set-aside programme whereby the government offers landholders incentives to enter into contracts to change the land use on a specified plot thereby providing ecosystem service benefits. It is administered by the Farm Service Authority (FSA), part of the United States Department of Agriculture (USDA), with support functions provided by the Natural Resource Conservation Service (NRCS), state forestry agencies, local soil and water conservation groups, and the private sector. It is funded by the government owned and operated Commodity Credit Corporation, created to support and protect farm income and prices. In 2010 USD 2 billion will be paid to secure retirement of 31 million acres of cropland. Over 80% of the CRP land is enrolled using a competitive bidding process, making the CRP the largest and longest running PES programme utilising inverse auctions. As such, there are valuable lessons to be learnt from the design and functioning of the CRP as it has evolved during the 23 years it has been in operation.

The CRP is not the only agri-environmental programme in the United States; it is part of a suite of incentive-based programmes targeting different aspects of the environment. This chapter focuses its analysis on the CRP because it is the dominant programme, but aspects of the other programmes are included where relevant. The chapter is organised as follows: Section 6.1 introduces the CRP in the context of other conservation programmes on agricultural land in the United States. Section 6.2 highlights important design elements of the CRP. Section 6.3 evaluates these design elements, including the use of inverse auctions, considering to what extent they contribute to the efficient functioning of the programme. Section 6.4 concludes, highlighting the design aspects which have contributed to the success of the CRP and lessons learned.

CRP context and objectives

Voluntary retirement programmes have been used in the United States to influence crop prices since the 1930's. However, the CRP, established by the 1985 Food Security Act, is the first cropland retirement programme explicitly following an environmental conservation agenda (Hellerstein and Hansen, 2009).

The CRP focuses on agricultural lands, the environmental impacts of which are diverse. For example, excess nitrogen loading in the Mississippi is the cause of eutrophication events which severely affect biodiversity in large areas of the Gulf of Mexico, termed the 'Dead Zone' (Rabalais *et al.,* 1997). Erosion, exacerbated by soil disturbance and the lack of vegetative cover, reduces the quality of agricultural land, forcing increased conversion of natural habitats. More than 80% of North American native grasslands have been lost since the mid 1800's (Samson and Knopf, 1994) leading to the rapid decline of grassland species. Wetland area in the United States has declined from about 221 million acres in the 1780's, to 103 million acres by mid-1980's (Dahl and Johnson, 1991). The downward trend continued through the 1990's, with losses averaging 31 000 acres per year between 1982 and 1992 (Heimlich *et al.*, 1998). Wetlands are particularly valuable biological resources because of their water purification functions, and their importance to many species for breeding, feeding, and shelter zones.

The environmental objectives of the CRP have evolved over time. In its initial form, the CRP targeted soil erosion reduction, although political support for the bill was bolstered by implications of reduced commodity surpluses. Additional conservation goals were included as the CRP was reauthorised in subsequent Farm Bills: protection of environmentally sensitive lands and improving water quality in the 1990 Farm Bill, and later enhancing wildlife habitat and improving air quality in the 1996 bill. These goals are achieved through retirement of cropland and the implementation of specified management practices. Reduced disturbance, decreased chemical run-offs, planting of grassland or tree cover, creation of wildlife corridors, habitat restoration, as well as the installation of grass filter-strips and riparian buffers, all contribute to protect highly erodible land, improve water quality and enhance wildlife habitat. In 2009 the CRP had over 30 million acres enrolled (Figure 6.1). The CRP is part of a portfolio of conservation projects which together tackle the environmental impacts of agriculture. To increase the effectiveness of the portfolio, each programme has specific aims, eligibility criteria, and payment mechanisms. The major programmes operating on agricultural land are outlined in Box 6.1.

Figure 6.1. **Acres of general CRP sign-up, 2009**

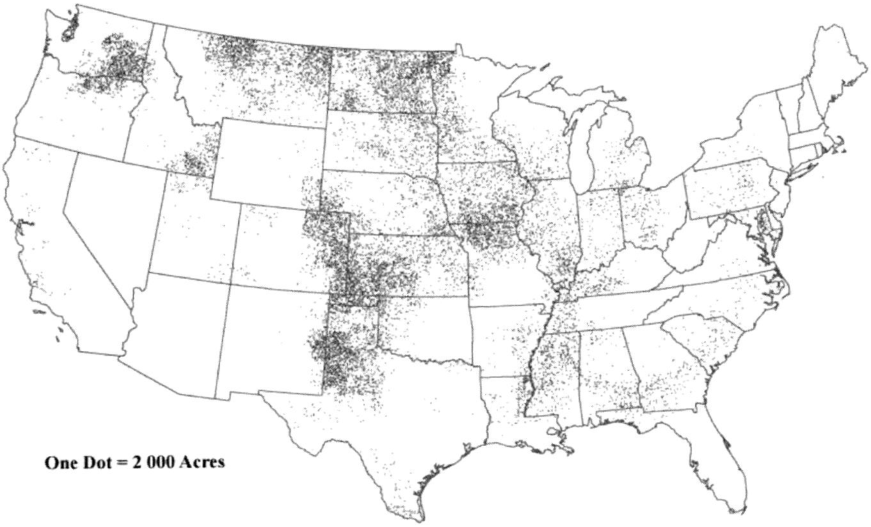

Source: ERS based on data from Farm Services Agency, USDA.

Box 6.1. **The USDA portfolio of conservation programmes**

In 2007 agricultural conservation spending represented about 16% of the USD 33.8 billion in Federal spending for natural resources and the environment (Claassen, 2009). The USDA uses PES and PES-like schemes to incentivise private investment in environmental stewardship, and increase the supply of ecosystem services from agricultural lands. Numerous conservation programmes with differing goals are in operation. The major ones are outlined below.

Land retirement programmes

- The Conservation Reserve Program: 10 to 15 year contracts for removing agricultural land from production to reduce soil erosion, improve water and air quality, and enhance wildlife habitat. The budget in 2010 will be USD 2 billion, about a third of all federal spending on conservation and recreation. The CRP consists of the following four sub-projects.

  1. General sign-up: auctioned contracts for whole field retirement, with implementation of various management practices. As of January 2010, there are 342 000 general sign-up contracts, representing 26.7 million acres of set aside land.

Box 6.1 continued over page

Box 6.1. **The USDA portfolio of conservation programmes** *(cont.)*

2. Continuous sign-up: non-competitive sign-up for partial field enrolment providing high quality environmental benefits through implementation of specific management practices. As of January 2010, there were 389 000 contracts, representing 4.4 million acres of set aside land.
3. Conservation Reserve Enhancement Program: Launched in 1997, as a subset of the continuous sign-up. Projects are initiated by local government, or non-government entities that identify an agriculture-related environmental issue of state or national significance. The project is then developed in coordination with the USDA tailoring the sign-up criteria to the local needs. Whole or part fields can be enrolled at anytime, receiving higher rental payments than the general CRP. Contracts contribute to the continuous CRP acreage and budget caps representing about 3.7% of the acreage and 9% of the payments in January 2010.
4. Farmable Wetlands Program: Pilot project fully integrated into the CRP 2008 to restore up to 1 million acres of farmable wetlands and associated buffers, to prevent continued degradation of wetland areas, improve water quality and prevent soil erosion, while providing valuable habitat for waterfowl and other wildlife. Contracts are enrolled via the continuous sign-up process; as of January 2010 there were 208 000 acres enrolled.

- The Wetlands Reserve Program (WRP): Authorised by the 1990 Food, Agriculture, Conservation and Trade Act to restore, protect, and enhance wetlands. Three types of contracts are available: Permanent Easement (representing 80% of contracts), 30-Year contracts, and Restoration Cost-Share Agreements. The acreage cap, which increased in 2008, is 3.041 million acres and sign-up is continuous.

Box 6.1 continued over page

Box 6.1. **The USDA portfolio of conservation programmes** (*cont.*)

Working land programmes

- Environmental Quality Incentives Program (EQIP): Launched in 1996, the EQIP provides farmers with assistance to improve environmental quality on farms. In some cases it may work in conjunction with local regulations. Between 2008 and 2012, 60% of the USD 7.25 billion budget is set aside for poultry and livestock, with the rest allocated for cropland programmes. Minimum contract length is one year, offering rental payments and up to 75% cost-share payments. Contracts are accepted on a continuous basis; however they are nonetheless ranked according to environmental benefits and economic costs. Demand for the EQIP is high. In 2007, for example, USD 993 million was assigned to contracts, however, the budget was an estimated USD 865 million short of the amount required to accept all offers. The high level of demand suggests competitive bidding may provide efficiency gains. Indeed prior to 2002, contracts were allocated using an inverse auction. As an indication of these gains, cost-share rates averaged 35% between 1996 and 2002, less than half of the 75% allowed. Moreover rental rates were, on average, 43% of the maximum rental rate (Cattaneo *et al.*, 2005).
- As part of the EQIP, Conservation Innovation Grants are available to local governments and non-for-profit organisations to stimulate the development of innovative conservation practices.
- Conservation Stewardship Program (CStP): The CStP replaces the Conservation Security Program following the 2008 Farm Bill, although existing contracts continue to be valid under the CStP. Landholders can enroll cropland, pasture, and non-industrial forest land. However, to be eligible landholders must have already addressed at least one resource concern throughout their farm, and agree to address at least one additional concern over the five year contract. The resource concerns relate to air, water and soil quality, as well as other aspects of environmental protection. The USDA aims to enroll 12.77 million acres per year, at an average cost of USD 18 per acre. Payments are dependent on the opportunity cost incurred by landholders and the expected environmental benefits.

6.2 The CRP general sign-up

USDA environmental programmes have traditionally used voluntary incentive-based approaches to conservation. The CRP is no exception, payments are offered to farmers to incentivise them to willingly change their land-use practices. However, the CRP is unique in that it incorporates an inverse auction into the contract selection process. The general sign-up represents 88% of the acreage under the CRP, and 75% of the payments; the remainder is allocated through a continuous sign-up process. In contrast, continuous sign-up is non-competitive, enrolling smaller areas of high quality land with sought after conservation potential (see Box 6.2).

Box 6.2. **The CRP continuous sign-up**

While general sign-up is used to enrol whole fields for retirement, the continuous sign-up focuses on small, high quality plots. It was initiated in 1996 and has since been expanded in 1997 and 2008. Landholders can enrol at any time through a non-competitive process; all eligible offers are accepted. Eligible offers propose the installation or restoration of riparian buffers, wildlife habitat buffers, wetland buffers, filter strips, grass waterways, shelterbelts, living snow fences, contour grass strips, salt tolerant vegetation, shallow water areas for wildlife, or may be any land within a pre-designated EPA public wellhead area. In general only a portion of the field is enrolled, but whole fields can be enrolled if more than 50% of the field is eligible, and when farming on the remainder is infeasible. Rental rates under the continuous sign-up are typically higher those of the general sign-up, with land in EPA-designated areas, and contracts offering more highly regarded management practices, receiving higher rental payments. Per-acre rental payments are higher for continuous sign-up partly due to the geographical location (there are a high percentage of sites are in the corn belt) and due to the greater incentives required to retire high quality, more productive, land in river and stream flood-plains. In addition, one-time sign-on incentives are available of up to USD 150 per acre, as well as initial cost-sharing which may be greater than 50%.

Eligibility

General sign-up auctions encourage eligible farmers to submit bids for 10 to 15 year contracts requiring the retirement of whole fields in return for annual rental payments. Supplementary payments are available for specific management practices, such as the installation of riparian buffers, and where initial costs are incurred, the USDA offers to share up to 50% of the cost. The use of land, and landholder, eligibility requirements are intended to ensure the environmental benefits of a contract are additional to the *status quo*. In other words, landholders should not submit lands which are either already in conservation use or would have been put to conservation use anyway. Producers must have owned or operated the land for at least 12 months prior to the close of the sign-up period, or must prove that the land was not acquired for the purpose of enrolling it in the CRP, for example through bequest. To be eligible the land must have been planted with an agricultural commodity for four of the six years prior to 2008 (the most recent Farm Bill), and must be physically and legally capable of being replanted.

The CRP general sign-up auction design

The general sign-up auction, administered by the USDA's FSA, is a single shot, sealed bid, discriminatory-price auction with a pricing cap (see Table 6.1). The auction evaluates bids based on cost and quality, aiming to select the most cost-effective contracts, and then compensating landholders for their individual opportunity costs. The cost-effectiveness of discriminatory-price auctions requires that a high level of competition is maintained. Competition reduces the ability of the landholders to exploit the information asymmetry associated with their opportunity costs, ensuring bids are as close to the landholders true opportunity costs as possible.

Table 6.1. **Key elements of the CRP general sign-up auction**

Issue	Key design element
Mechanism	Inverse auction; single shot (bidders cannot revise their bids), sealed bid (bidders cannot view competitors bids), discriminatory-price auction (successful bidders are paid their bid price).
Price	Successful bidders are paid their bid price in differentiated payments. Supplementary fixed payments for specific management practices. Optional cost-share payments of up to 50% of initial implementation costs.
Bids	Sealed bids, which include information on the environmental quality of the land, proposed management practices, requested PES payment, and the amount of cost-share requested.
Rounds	Sequential auctions held over extended period of time.
Bid selection	Based on Environmental Benefit Index, which includes costs evaluation.
Selection cut-off	Pricing cap set for each bidder, depending on local land rental rates and bid specific soil productivity rating. The price caps are revealed to bidders.
Decision-making	Local FSA offices select eligible bids; National FSA select winning bids.
Payments	Annual rental payments. Cost-share payments are made when practices are installed.
Ongoing monitoring, reporting and evaluation	Local NRCS offices undertake compliance review.

Source: OECD, 2010.

In the general sign-up auctions, landholders submit their bids, including environmental information on the plot and their proposed management practices, as well as the requested contract payment.[3] The USDA ranks the bids according to potential environmental benefits and cost, incorporating this information into an Environmental Benefit Index (EBI). This index was introduced in 1991 allowing quantifiable assessment of the potential conservation outcomes, such that the contracts offering the highest benefits for least cost can be selected. Implicit in its design is the trade-off between the different environmental aims (see Section 6.3). Indeed since its inception the details of the EBI have changed as conservation priorities have changed. Currently wildlife, water quality and local erosion control benefits each carry a maximum of 100 points; up to 50 points are available for benefits enduring past contract expiration; 45 points for air quality benefits; and up to 150 points for relative cost (see Box 6.3).

The EBI contains some elements which are out of the bidders' control, inherent to the quality of the land on offer. However, landholders can make their bids more attractive by offering the implementation of high value management practices and increasing cost reductions. Points can be gained from cost reductions by forgoing the cost-sharing payment or reducing the requested annual rental rate. Competition for contracts is national, *i.e.* all the bids from different states are pooled and contracts with the highest EBI score selected.

Prior to submitting a bid, landholders are informed of the maximum acceptable per acre rental rate the USDA is willing to pay. It is calculated using the county average cropland rental rates, and the relative productivity of the dominant soil types within each plot. Using market information to set the maximum rate ensures that the payments are reasonably close to the landholders opportunity cost from not producing on the land, and avoids unreasonably high bids. The maximum rate effectively acts like a pricing cap for the retirement contracts.

Enforcing contracts

Once enrolled, the farmer is under a legal obligation to carry out the management practices as stipulated in the contract. The incentive to do so require payments to continue to cover the opportunity costs of participation for the duration of the contract. If they fail to do so, subject to rising crop revenues or a miscalculation by the landholder prior to submitting the bid, for example, the landholders have an incentive to breach the contract. The closer the payment is to the landholders' minimum willingness to accept (WTA), the more susceptible it is to changes in opportunity costs. This makes effective enforcement even more important when using auctions.

Box 6.3. **The CRP Environmental Benefits Index**

The breakdown of how the EBI points are allocated between different environmental benefits is outlined here. The final EBI score is the sum of the individual scores for the following six factors.

1. **Wildlife Factor Benefits** – up to 100 points.
   - Wildlife habitat cover benefit; 0 to 50 points, awarded for different planting mixtures.
   - Wildlife enhancement; 0, 5, or 20 points, awarded for specific practices likely to increase biodiversity benefits.
   - Wildlife priority areas; 0 or 30 points, awarded for contracts within conservation priority areas, as designated by FSA.
2. **Water Quality Benefits** – up to 100 points.
   - Location; 0 or 30 points, awarded for contracts within priority areas, where water quality is impaired by crop production.
   - Groundwater quality; 0 to 25 points, dependent on soil type, the potential leaching of pesticides and nutrients into groundwater, and the population impacted.
   - Surface water quality; 0 to 45 points, awarded depending on runoff and waterway sedimentation potential, and the relative level of surface water impairment in the watershed.
3. **Erosion Factor** – up to 100 points.
   - Erosion factor; 0 to 100 points, awarded dependent on the potential for on-site erosion to decrease the long-term productivity of the land, as measured using an Erodability Index.
4. **Enduring Benefits Factor** – up to 50 points.
   - Enduring benefits factor; 0 to 50 points, awarded for contracts providing benefits that are likely to endure beyond the contract period.
5. **Air Quality Benefits** – up to 45 points.
   - Wind erosion impacts; 0 to 25 points, awarded depending on the Erodability Index, calculated from the biophysical attributes of the land, and the population impacted by airborne particulates.
   - Wind erosion soils list; 0 or 5 points, awarded for land with particularly sensitive soils or damaging particles (dominantly organic or volcanic).
   - Air quality zones; 0 or 5 points, awarded for contracts with high erodability potential and that are located within designated priority areas.

Box 6.3 continued over page

Box 6.3. **The CRP Environmental Benefits Index** *(cont.)*

- Carbon sequestration; 3 to 10 points, awarded after evaluation of the benefits from greenhouse gas sequestration over the life of the contract.

6. **Cost** – the number of available points is determined by the USDA after the bidding process is complete.
   - Forgoing cost-share; 0 or 10 points, all projects that include cost-share receive 0 points.
   - Rental reductions; 0 to 15 points, bids are awarded one point for each dollar discount from the maximum rental rate, discounts over USD 15 all receive 15 points.
   - In addition, points are assigned depending on the cost of the project, relative to the highest national maximum rental rate. The number of points is subject to the choice of W in the total cost points equation below. Since sign-up 16 in 1997 its value has been set at 125, such that a total of 150 points are available for cost, reduced from 200 in previous years.
   - Total cost points is therefore given by:

     $C = w\,(1 - r/H) + 10\,(1 - s) + \min(15, r^m - r)$,
   - where, C is cost points, W is an arbitrary value set by the USDA after bids are received, r is the proposed rental rate, r^m is maximum rental rate for the parcel being offered (which is a function of country average rental rates and the soil type(s) prevalent on the parcel), H is the highest national maximum rental rate, and s is the share cost decision (1 share, 0 not).

Source: USDA (2006).

If a landholder wishes to exit a contract early there are provisions to do so, at a cost. The landholder must refund the rental and cost-share payments in full plus interest. Compliance enforcement issues are handled on a case-by-case basis. A spot-check is conducted on less than 1% of CRP farms annually. It is left up to the individual counties and States if they want to do additional compliance checking. In 2007, for example, 808 landholders where randomly selected for spot-checks (from a population of over 450 000) with about 1% found to be non-compliant.

6.3 The CRP environmental and cost effectiveness

The environmental effectiveness of the CRP

The environmental effectiveness of the CRP is dependent on the supply of the desired ecosystem services, prioritised in the EBI. These services must be additional to what would have been provided in the absence of the programme and the service provided must be appropriate for the natural context.

Environmental benefits

Since the 1990's the CRP has maintained over 30 million acres of land enrolled. Initially, the accomplishments of the CRP were stated in terms of area of land retired or wetlands restored. In 2000, for example, an enrolment target of 24 million acres of highly erodible land was set, with 23.7 million acres enrolled. However, area based assessments do not provide a representative view of the real environmental outcomes and benefits of the programme. A comprehensive evaluation of the CRP requires an assessment of the extent to which the specific conservation aims of the programme have been achieved. In addition to enrolled land area, the resultant ecological impacts arising from reduced pesticide run-off, or the installation of riparian buffers, for example, needs to be examined to infer the biodiversity benefits. Since 2000 more detailed indicators have been employed to quantify the CRP performance. In 2003, a target of 447 million tons of avoided soil erosion was set (and achieved). The transition from area based targets to output targets illustrates the increased use of quantifiable performance indictors to evaluate the CRP benefits. However, the USDA acknowledges that these indicators are still not an adequate way of accurately communicating the real conservation benefits (Hyberg, 2004). Thus, prompted by an 80% increase in funding for conservation programmes between the 1996 Farm Bill and the 2002 bill, the Conservation Effects Assessment Program (CEAP) was launched to assess the environmental performance of conservation practices, including the CRP, across the United States.

The CEAP is a joint project between the Natural Resources Conservation Service (NRCS) and the Agricultural Research Service (ARS). The CEAP aims to give a scientifically credible assessment of the national environmental benefits obtained from USDA conservation programmes. Published results have demonstrated the substantial benefits to local freshwater and grassland ecosystems. It may be several years before the CEAP publishes national level conclusions; however, there are a number of interesting preliminary results for consideration. A selection of results is presented in Box 6.4, using the Prairie Pothole Region as an example.

Box 6.4. **Ecosystem services derived from wetland conservation in the Prairie Pothole Region**

Preliminary results from the Conservation Effects Assessment Project (CEAP)

The Prairie Pothole Region covers an area of over 220 million acres extending from the north-central American Great Plains to south-central Canada. It is typically dominated by mid-, to tall-grass lands, containing thousands of shallow wetlands, known as potholes. This habitat supports more than 50% of the United States' migratory waterfowl (US EPA, 2008). Between the 1780's and 1980's however, huge expanses of wetlands were drained to be used in agriculture, Iowa, for example, lost 98% of its prairie land (Dahl, 1990). Currently more than 7 million acres are enrolled through the CRP and WRP programmes.

Gleason *et al.* (2008a) evaluate the plant communities, carbon sequestration, sediment and nutrient loading, as well as the wildlife habitat potential, associated with these conservation efforts. The study examined temporary, seasonal and semi-permanent wetlands, covering an alteration gradient from highly altered, to minimally altered, allowing the benefits of managed lands to be compared to native wetlands and cropland.

Plant community quality and richness was assessed using an index of floristic quality and species richness[4] (Laubhan and Gleason, 2008). The results indicated that restored catchments had a significantly higher index value than cropped catchments, but a lower value than that of native prairie catchments.

No significant difference was found between soil organic carbon (SOC) levels in cropped and restored wetlands, highlighting the fragility of the microbial soil community. Again, however, as the sites mature the sequestration benefits may increase (Gleason *et al.,* 2008b).

Sedimentation and nutrient run-off from upland cropland is a major cause of degradation to the adjacent wetlands (Tangen and Gleason, 2008). The conversion of 680 000 acres of enrolled uplands reduces total soil loss by nearly 2 million tons per year. For the same area, it is estimated that nitrogen and phosphorus losses are reduced by 5.6 thousand tons per year, and 75 tons per year, respectively, significantly improving the environmental quality of the low lying wetlands, and avoiding the loss of potential productivity of the uplands.

Box 6.4 continued over page

Box 6.4. **Ecosystem services derived from wetland conservation in the Prairie Pothole Region**

Preliminary results from the Conservation Effects Assessment Project (CEAP)

(cont.)

Wildlife habitat potential was assessed for area-sensitive bird species, based on their habitat requisites, and the spatial and structural nature of the site (Laubhan *et al.,* 2008). The survival and reproduction of many species is highly dependent on these habitat attributes, and has been adversely affected by the fragmented distribution of the remaining native habitat. The results showed that both the grasslands and the wetlands provided adequate habitat for the species evaluated. Adair and James (2004) support this conclusion, reviewing original studies of avian populations in this area, quoting the positive effects on songbirds and waterfowl. It was estimated that CRP lands in N. Dakota, S. Dakota, and north-eastern Montana led to an increase in waterfowl populations (mallard, gadwall, blue-winged teal, northern shoveler and northern pintail) of 2 million ducks per year between 1992 and 2004, representing a 30% increase in productivity compared with same area in the absence of CRP cover (Reynolds *et al.,* 2004). Furthermore, Johnson and Igl (1995) predicted that populations of at least five species of songbirds in North Dakota would decline by 17% or more if CRP plots were replaced by cropland. Songbirds are in decline in the United States, requiring extensive, densely vegetated grasslands. The CRP has successfully tempered declines that otherwise would have led to increases in the number of endangered or threatened species.

Additionality and leakage

To attribute the environmental benefits achieved to the CRP, the land use changes must be additional to what would have happened anyway. Equally, the retirement of a plot of land must not have motivated the subsequent conversion of natural land to cropland in another area. This is the leakage problem, or slippage as it is often referred to in the United States programmes.

An assessment by Lubowski *et al.* (2003) estimated that about 15% of the land enrolled in the CRP would have shifted from crop-use anyway. However, this includes conversion to grazing and forestry, the environmental benefits from which would not necessarily be the same.

The degree of additionality can also be assessed when contracts are re-enrolled. This is because additional benefits are only gained from re-enrolment if landholders would have returned the land to agricultural uses without re-enrolment. Sullivan *et al.* (2004) evaluate the changes in land use following the withdrawal of 3.6 million acres from the CRP in 1997; 63% returned the land to crop production, 31% to pasture or rangeland, and the remaining 6% kept the land in non-farm uses. However, these decisions were made voluntarily and so cannot be used to predict changes if re-enrolment was disallowed. To this end, Sullivan *et al.*, model landholder decisions, estimating that 51% of CRP land would be returned to crop production in the absence of CRP payments. Land planted with trees was less likely to be converted, and the decision making process was heavily influenced by the potential profitability of the land, suggesting increases in crop revenues might encourage more landholders to revert the land use to produce crops. An important consequence of bringing CRP land back into production is that many of the environmental benefits obtained over the course of the contract are quickly lost, for example the soil organic carbon (SOC) would be rapidly released to the atmosphere, and the wildlife population would revert to previous levels following the reduction in their habitat. Therefore, there is a valid argument for prioritising re-enrolment of expiring contracts over new enrolment to avoid these losses.

As an indication of leakage, Wu (2000) noted that by 1992, 17.63 million acres of cropland had been retired in the Corn Belt, Lake States and Northern Plain, but that total cropland acres were only reduced by 13.69 million acres. At a glance this might suggest leakage is an issue, however, these changes can also be explained by the re-introduction of land enrolled from the conclusion of other land retirement programmes, which dominated the CRP in terms of acres enrolled until 1990 (Hellerstein and Hansen, 2009).

Estimating the extent to which leakage occurs is a difficult empirical problem because the current situation must be compared to a scenario without the programme. The incentives to bring natural land into production will be based on the price effect associated with reduced supply and the landholders' substitution effects. Wu (2000) modelled these incentives, estimating that for every 100 acres retired, 20 acres is brought into production. However, using the same data set, Roberts and Bucholz (2005) question Wu's methodology, suggesting that leakage is only negligible.

To dis-incentivise landholders from bringing natural land into production, a 'sodsaver' provision was included in the 2008 Farm Bill. This removes federal support for newly converted land; the land would be ineligible for all support programmes including marketing assistance loads, disaster relief and insurance payments. The provision is voluntary on a state by state basis, but to date no State has implemented it.

Environmental shortcomings

While the environmental benefits of the CRP have been widely acknowledged, some concern has been raised over instances of negative environmental effects of the CRP. Natural ecosystems are characterised by a range of habitats at different stages of ecosystem succession, providing niches for a community of diverse species. To achieve the maximum environmental benefit from reverting land use from cropland to conservation land, it is important to acknowledge the subtleties of the natural system. Bidwell and Engle (2004) highlight one of the main shortcomings of the CRP as being the lack of contextual relevance of the conservation practices to the local needs of habitat specialists.

For example, in prairie lands, the planting of mid- and tall-grasses on areas historically dominated by short-grasses decreases the habitat value for species with a habitat niche limited to short prairie, such as the Mountain Plover. The planting, or unchecked invasion, of woody shrubs and trees in prairie lands is particularly damaging because it attracts habitat generalists, such as White-Tailed Deer, Raccoon, and Brown-Headed Cowbird, from the adjacent forests. These are formidable competitors and predators to native species.

This demonstrates the importance of implementing the proper management practices to native wildlife. Furthermore, Bidwell and Engle note the influence of spatial distribution on the potential environmental benefits of CRP plots; numerous highly fragmented plots often fail to provide significant benefits, compared to the same area distributed in a few large tracts. These issues concern how the details of potential contracts are evaluated and selected by the EBI; if the index fails to select the contracts proposing relevant management practices the resultant outcome can have adverse effects on the natural ecosystem.

Bid evaluation

The EBI was introduced following the inclusion of diverse conservation goals as a way of evaluating and selecting bids in an efficient manner. The EBI evaluates both the environmental quality and cost effectiveness simultaneously. The broadening of the goals of the CRP resulted in increases in the amount of potentially eligible land from 100 million acres in 1986 to 240 million acres in 1997[5] (Osborn, 1997). This was associated with an increase in competition for contracts and a decline in the average rental payment from USD 50 per acre to USD 39 per acre, with a greater proportion of landholders offering discounts on the maximum rental rate. Pooling landholders offering different benefits may increase competition, however it also has an effect on the ability of the project to target specific environmental concerns. The formulation of the CRP at present, giving equal weight to the major benefit categories (water quality, erosion reduction, and wildlife benefits) favours a generalised approach.

Insights into the relative benefits of the categories constituting the EBI can be gained by examining the share of points awarded for the different benefit categories within accepted contracts, depicted in Figure 6.2 for sign-ups between 1997 and 2003. Rarely did a single environmental factor account for more than 40% of contract points, emphasising the generality of the benefits on the selected land. On average, wildlife habitat accounts for about 20% of the EBI score; water quality for 16%, and on-site erodability for 19%. Cost is the dominant factor, accounting for an average of 35% of EBI points, and more than 40% in a quarter of contracts (Claassen *et al.*, 2008).

Figure 6.2. **The relative share of points awarded by category within all accepted bids, CRP general sign-up 1997 to 2003**

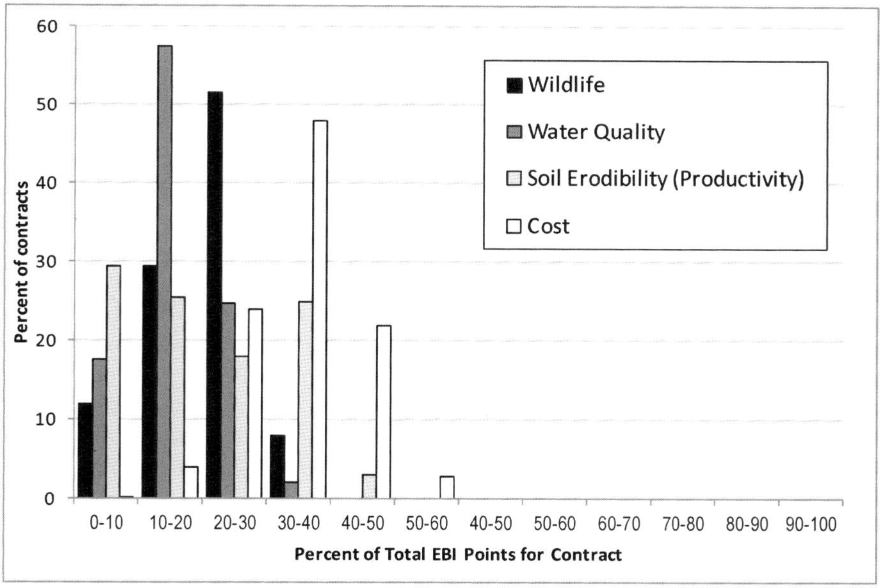

Source: Claassen *et al.*, 2008.

There are trade-offs involved in the design of the EBI targeting mechanism. By targeting general benefits the EBI discriminates against sites offering exceptional benefits in one category, but few benefits in other categories, irrespective of locally specific resource concerns. On the other hand, a more specific targeting mechanism may result in omitting sites that have high aggregate benefits, but do not excel in any one dimension. Analysis by the Soil and Water Conservation Society and the Environmental Defence Fund (2008) suggested that improvement to the EBI could be made to avoid enrolling 'mediocre' sites by increasing the points difference awarded to high and low quality applications within each benefit category. Moreover, they suggest the inclusion of location specific management practices, and modifying the EBI category point weightings by location. This would ensure the contracts offering the appropriate management practices are enrolled within each location. Furthermore, they propose the use of Geographic Information Systems (GIS) data to evaluate the spatial nature of bids, giving greater priority to those adjacent to existing conservation lands. Complementary conservation programmes such as the continuous sign-up and the EQIP offset some of the generalities of the EBI targeting mechanism by focusing on specific high quality sites, considering local and regional environmental priorities.

The cost-effectiveness of the CRP

Maximising benefits per cost

Cost effectiveness requires the CRP to select the contracts with the highest per cost environmental benefits, via the EBI. Prior to 1997 the EBI was calculated with purely environmental data and the final score divided by the contract cost (EBI/USD). In this system it could be easily verified that the maximum gains per dollar were secured. The problem being however, that the final score was highly dependent on the local maximum rental rate, because bids are anchored to the local maximum. Areas with high rental rates (implying highly productive lands) were thus discriminated against. In the current system, contract cost is incorporated by allocating it a quantity of points, which go towards the final EBI point total. This corrects for the bias, however, the drawback is that it makes it more difficult to assess if funds are used in the most cost-effective manner, because the relative importance of cost versus the different environmental benefit categories has to be decided.

A complimentary analysis of EBI cost-effectiveness focuses on the environmental benefits from each category in terms of monetary value. The most efficient EBI would then prioritise the environmental category offering the highest net marginal value. Awarding the environmental categories equal weight implies that their benefits are assumed to have equal net marginal value, which is unlikely to be the case.

To investigate how the EBI could improve targeting, Feather *et al.* (1999) carried out a nonmarket economic valuation of freshwater-based recreation, wildlife viewing, and pheasant hunting benefits reaped in 1992. The results indicate that the value of wildlife viewing (USD 10.02 per acre) far exceeds that of pheasant hunting (USD 2.36 per acre) and freshwater recreation (USD 1.07 per acre), suggesting potential cost-effectiveness gains could be gained if the EBI was re-prioritised in favour of wildlife habitat. To illustrate this, a simulation using real bid data was run. EBI scores were recalculated according to the adjusted index and the potential environmental benefit values calculated. Total water-based recreation benefits increased by 255%, and wildlife viewing benefits by 83%, while pheasant hunting benefits decreased 13%. These benefits were not evenly distributed across the country, and thus could be further increased with the use of locally specific EBI's. The analysis is not complete, but it illustrates how the EBI could be used to adjust targeting and cost effectiveness.

Claassen *et al.* (2008) also note that farmers already have a private incentive to maintain soil productivity on their land so the points allocated to on-site erosion benefits (100) are misplaced.

The use of auctions to improve cost effectiveness

Competitive auctions are incorporated into the general sign-up process of the CRP as a tool to improve the ability of the regulator to obtain maximum environmental benefits from a given budget. Cost effectiveness requires that the payments to landholders are equal, or close to their minimum WTA to forgo income from producing on their land. All else being equal, their opportunity costs from lost income should equal their minimum WTA. However, the information asymmetry of the potential income loss between the landholders and the regulator gives the landholders an incentive to inflate their bids above their minimum WTA. The competitive nature of the auction reduces the landholders' extractable information rents, forcing them to trade-off the risk of losing the contract with the potential to reap higher rental payments. Bidders can make their bids more attractive by offering high quality additional management practices, rental discounts from the maximum rental rate, and by forgoing cost-sharing.

The CRP utilises a discriminative price auction. Provided sufficient competition, discriminative price auctions are efficient because the differentiated payments set the price for each contract according to individual opportunity costs, maximising the purchasable benefits for a fixed budget. To maintain competition the auction should be designed to minimise the bidders' knowledge of the buyer's preferences, in terms of benefits provided and willingness to pay, and their information on the characteristics of their competitors. This section considers how effectively the CRP auction maintains competition, and thus how cost-effective the use of auctions is in allocating contracts.

In an analysis of the difference between the landholders' opportunity costs and the received rental payments for two sign-up auctions in 1999 and 2003, Kirwan *et al.* (2005) estimate that payments are 10 to 40% above the minimum necessary to cover the lost farming income. Although this appears to imply there are large inefficiencies, it does not necessarily mean that 10 to 40% of the payments are lost to information rents. Kirwan *et al.*, recognise that this may simply reflect the premium necessary to encourage farmers to change their habits, encompassing landholder transaction costs, compensation for lost land use options, and the amount required to reveal their private cost information. These additional elements mean the landholder minimum WTA may in fact be greater than just the opportunity costs of lost farming income.

However, the efficiency of an auction is dependent on competition, requiring a large number of bidders with heterogeneous costs. If competition is weak, bidders have less incentive to offer discounts on the maximum rental rate, or forgo cost-sharing, because the risk of losing the contract are lower, allowing bidders to inflate their bids above their minimum WTA. Analysis of the bids received and accepted for five auctions between 1997 and 2003 reveals that competition was not especially intense; in the first four auctions 65-75% of bids were accepted, with 50% accepted in the 5th auction in 2003. The proportion of bids with discounts offered also declined across auctions. It is therefore likely that annual rental payments are not perfectly in line with landholder minimum WTA (Claassen *et al.*, 2008).

The use of a revealed maximum rental rate, effectively a contract price cap, has an important effect on competition and cost effectiveness. The cap is set using costly-to-fake information available to the regulator about the potential opportunity costs, and is revealed to bidders in advance of the auction. This is effective in avoiding unreasonably high bids, and increases transparency for participants. The cap also minimises price inflation in the land rental market, because if the CRP paid above market rental rates it could cause these prices to increase, affecting the wider economy. However, there are a number of bidding implications of the cap. Firstly, as the cap is revealed it informs the bidders of the buyer's willingness to pay, and can act as a pricing anchor for bids. When evaluating their WTA, landholders will formulate their price based on the cap, which may introduce a systematic judgment bias. Bids will therefore be clustered closer to the cap than may have otherwise have been the case. This is exacerbated by the fact that the awarding of additional cost points increases for discounts up to USD 15, after which point they are constant. Anchoring thus reduces the ability of the regulator to differentiate between bids and may potentially reduce the cost-effectiveness of the resultant selection.

Secondly, the revealed cap can reduce the incentives of landholders with especially high quality land to implement additional improvement because they are confident their bid is still attractive to the regulator at the maximum price. Offering additional improvements would value the contract above this price, but they have limited incentives to do so because they will incur higher costs without the corresponding compensation. Bids with high inherent EBI scores[6] are thus found to demand the maximum rental rate, and offer few additional benefits, while bids with low inherent EBI scores generally try to improve their bid by offering discounts or additional improvements (Claassen *et al.,* 2008; Islik, 2005). The choice to include a price cap is therefore an outcome of trading off potential programme cost-effectiveness reductions with the broader political and socio-economic concerns. This highlights the importance of considering the wider context of PES programmes during their design.

Fundamental in the choice to use auctions over a fixed price scheme is that the cost-effectiveness gains from auctioning, less the additional transaction costs from implementing a more complex programme, are greater than the losses of a fixed price scheme. To assess the cost-effectiveness of the CRP auctions, information on transaction costs incurred is required. Transaction costs encompass the costs of designing the programme, the landholders costs of submitting an application and the regulators costs of processing applications, selecting participants, entering into contracts, making payments, monitoring compliance, and enforcement activities. Initial costs of researching, designing and setting up the programme are likely to be significant; however, the costs are dissipated throughout the lifespan of the programme (currently in its 24th year). In 2004 USD 530 million was spent on ongoing research projects and data collection. Recurrent operational costs can be estimated from the reported USDA's FSA salaries and expenses of USD 15.5 million in 2004, less than 1% of the CRP expenditure.

Theory dictates that auctions are a more efficient way of allocating contracts and targeting conservation efforts. There are a number of variables in the design of auctions (see Chapter 1) which will affect to what extent they reduce the information rents extracted by landholders, and despite the caveats of some of the elements of the CRP auction design, highlighted above, in general the CRP auctions appear to be effective (Claassen *et al.,* 2008).

The decision of whether to allocate contracts using auctions, or without competition, in the general, and continuous sign-up, respectively, reflects the different aims of these two complementary components of the CRP. The general sign-up enlists large areas of set-aside land providing (principally) in-situ benefits; such sites are relatively common and so an auction can be used to discriminate between them, while improving cost effectiveness. In contrast, the continuous sign-up enrols small plots with high quality benefits that will provide environmental services for a wider area. These sites are not only of higher value but are comparatively scarce, such that ensuring these benefits are captured takes precedence over cost effectiveness. In this way the USDA attempts to target its conservation projects on specific environmental issues to increase their impact.

The cost-effectiveness of the USDA's portfolio of conservation projects as a whole requires that they continually address the conservation concerns with the lowest net marginal cost[7] at any point in time. Since the launch of the CRP, the conservation focus has changed accordingly, not only within the CRP, but between the different programmes. Since 2002 there has been a shift in emphasis from land retirement programmes, such as the CRP and WRP, towards working land conservation programmes. The 2008 Farm Bill re-enforced this policy with average annual funding increases for working lands programmes up from USD 1.05 billion between 2002 and 2007, to USD 2.34 billion between 2008 and 2012, making the funding for the EQIP and CStP greater than that for the CRP, which has traditionally been the dominant programme (Figure 6.3). Moreover, the acreage cap for the CRP was decreased from 39.2 million acres in 2002 to 32 million acres from 2009. A possible rational for this shift in policy is that conservation benefits from working lands are now considered to have a lower net marginal cost relative to the remaining conservation benefits available from increasing land retirement. This is no-doubt emphasised by the recent increases in crop prices inflating the economic burden of land retirement. The cost of operating these programmes could perhaps be reduced further by increasing the use of competitive auctions.

Government intervention through programmes such as the CRP is intended to increase social welfare. Concerns have been raised that the CRP may have contributed to rural population declines and reducing the agricultural economy.[8] However, a thorough analysis by Sullivan *et al.* (2004) suggests that increases in recreational activities dissipated any negative effects. Moreover, attempts to monetise the environmental, social, and industrial[9] benefits reveal that the total economic benefits of the CRP are likely to offset any economic costs (Bangsund *et al.,* 2003; Feather *et al.,* 1999; Ribaudo, 1986; Ribaudo *et al.,* 1990).

Figure 6.3. **Trends in USDA agri-environmental expenditures**

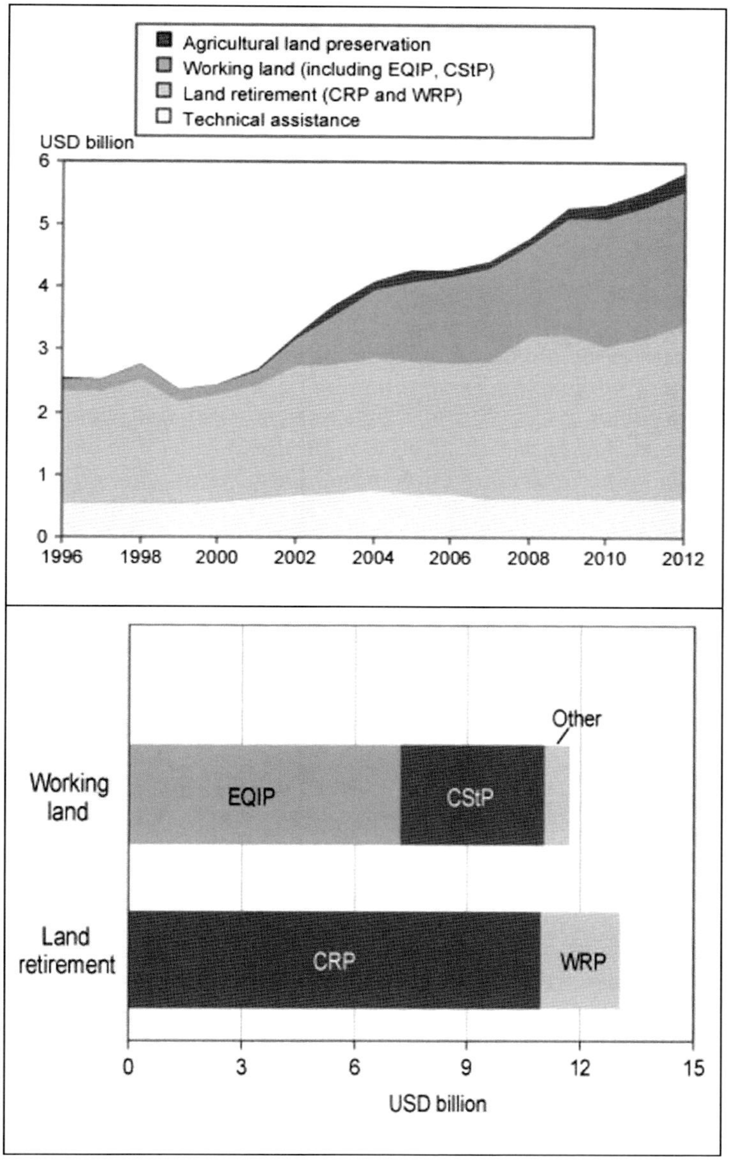

Source: Claassen, 2009.

6.4 Conclusions

The Conservation Reserve Program has much to offer in terms of experience in the design and the implementation of inverse auctions in PES programmes. The general sign-up utilises a competitive inverse auction combined with an EBI to evaluate contracts in terms of environmental quality and cost. The auction is a single shot, sealed bid, discriminative price auction with a pricing cap. This mechanism is considered to yield considerable cost-effectiveness gains over an alternative uniform price scheme. Nevertheless there are some design elements which have been criticised. In particular, the use of a revealed pricing cap which is disclosed to bidders may result in a reduced spread of payment bids, limiting the effectiveness of the bid evaluation process. Moreover, it limits the incentives of high quality landholders to furnish further improvements, or even participate. This issue is to some extent rectified by the use of complimentary programmes aimed at high quality land. The extended use of auctions in some of these programmes has increased cost effectiveness in the past. The use of auctions more widely in USDA programmes could further increase the cost-effectiveness of the USDA conservation portfolio.

The CRP has responded to changing priorities, modifying its goals over the years to reflect the changing environment within which it functions. The development of the EBI in 1991, the inclusion of a continuous sign-up in 1996, and the shift in emphasis to working lands conservation in 2002, are a few examples of this.

The size and scope of the CRP is perhaps one of its biggest challenges. The United States is home to highly heterogeneous environments, with contrasting conservation priorities. Improving the location specificity of the CRP management practices, together with ensuring proper implementation, will be important issues for the CRP going forward to secure the maximum potential environmental benefits are obtained from the programme. In 2010 and 2011, contracts representing 9.17 million acres are due to expire. To ensure the future of the CRP, the payments must continue to be competitive against the backdrop of rising crop demand and revenues.

Notes

1. The United States has 940 million acres of grazing and crop land, covering 52% of the land area (USDA, 2002).
2. Critically Endangered, Endangered and Vulnerable.

3. The bid price is effectively a combination of the required annual land rental rate and whether cost-share assistance is requested. A farmer's willingness to accept a contract is dependent on these two payments.

4. Floristic quality index, used to assess habitat management efforts; each species in a region is assigned a score (0-10) based on their tolerance to disturbance and site fidelity, low tolerance and high fidelity receiving a greater score (crops and non-native species receiving a score of 0).

   Species richness, used to measure species diversity in a given area; in this case simply the number of species found as proportion of regional total, diversity increases as score approaches 1.

   The index total is given by the product of floristic quality and species richness.

5. As eligibility has changed little since 1997, this figure will be more or less the same today.

6. Inherent EBI score refers to the exogenous EBI value of the land, the EBI attained by minimal management practices, with no price discount and accepting cost-sharing.

7. Those for which a given environmental gain are achieved at least cost.

8. The Conservation Reserve Program, Proceedings of a National Conference, 2004.

9. For example, from reduced water purification and de-sedimentation.

References

Adair S. and B. James (2004), "The Conservation Reserve Program: Proven benefits in the Prairie Pothole Region", in A. Allen and M. Vandever (eds.), *The CRP – planting for the future: proceedings of a National Conference*, Colorado, June 6–9, U.S. Geological Survey.

Bangsund D., N. Hodur and L. Leistritz (2004), "Agricultural and recreational impacts of the Conservation Reserve Program in rural North Dakota, USA", *Journal of Environmental Management*, Vol. 71.

Bidwell T. and D. Engle (2004), "Fine tuning the Conservation Reserve Program for biological diversity and native wildlife", in A. Allen and M. Vandever (eds.), *The CRP – planting for the future: proceedings of a National Conference*, Colorado, June 6–9, U.S. Geological Survey.

Cattaneo A., R. Claassen, R. Johansson and M. Weinberg (2005), *Flexible conservation measure on working land: what challenges lie ahead?*, USDA Economic Research Service, Report No. ERR-5.

Cattaneo, A. (2003), "The pursuit of efficiency and its unintended consequences: contract withdrawals in the environmental quality incentives program", *Review of Agricultural Economics*, Vol. 2.

Claassen R. (2009), USDA Briefing Room – Conservation Policy: Background. http://www.ers.usda.gov/briefing/conservationpolicy/background.htm.

Claassen R., A. Cattaneo and R. Johansson (2008), "Cost-effective design of agri-environmental payment programs: U.S. experience in theory and practice", *Ecological Economics*, Vol. 65.

Dahl T. (1990), "Wetland losses in the United States 1780's to 1980's", U.S. Fish and Wildlife Service, Washington DC.

Dahl T. and C. Johnson (1991), "Status and trends of wetlands in the conterminous United States, mid-1970s to mid-1980s", U. S. Fish and Wildlife Service, Washington DC.

Feather P., D. Hellerstein and L. Hansen (1999), "Economic valuation of environmental benefits and the targeting of conservation programs: the case of the CRP", Agricultural Economic Report No. 778, USDA Economic Research Service.

Gleason R., M. Laubhan and N. Euliss (2008a), "The United States Prairie Pothole Region with an emphasis on the U.S. Department of Agriculture Conservation Reserve and Wetlands Reserve Programs", United States Geological Survey.

Gleason R., B. Tangen and M. Laubhan (2008b), "Carbon Sequestration", in R. Gleason, M. Laubhan and N. Euliss (2008), "Ecosystem services derived from wetland conservation practices in The United States Prairie Pothole Region with an emphasis on the U.S. Department of Agriculture Conservation Reserve and Wetlands Reserve Programs", United States Geological Survey.

Heimlich R., K. Wiebe, R. Claassen, D. Gadsby and R. House (1998), "Wetlands and Agriculture: Private Interests and Public Benefits", USDA, Economic Research Service.

Hellerstein D. and L. Hansen (2009), USDA Briefing Room – Conservation Policy: Land Retirement Programs. http://www.usda.gov.

Hyberg S. (2004), "The role of science in guiding the conservation reserve program: past and future. USDA", in A. Allen and M. Vandever (eds.),

The CRP – planting for the future: proceedings of a National Conference, Colorado, June 6–9, United States Geological Survey.

Isik, M. (2005), "A theoretical and empirical analysis of Conservation Reserve Program participation under uncertainty", Presented at, American Agricultural Economics Association Annual Meeting, Rhode Island, July 24–5, 2005.

IUCN (2009a), *Wildlife in a changing world: An analysis of the 2008 IUCN Red List of Threatened Species*, Gland, Switzerland, IUCN.

IUCN (2009b), *IUCN Red List of Threatened Species.* Version 2009. http://www.iucnredlist.org.

Johnson D. and L. Igl (1995), "Contributions of the Conservation Reserve Program to populations of breeding birds in North Dakota", *Wilson Bulletin*, Vol. 107, No. 4.

Kirwan B., R. Lubowski and M. Roberts (2005), "How cost effective are land retirement auctions? Estimating the difference between payments and willingness to accept in the Conservation Reserve Program", *American Journal of Agricultural Economics*, Vol. 87.

Laubhan M. and R. Gleason (2008), "Plant community quality and richness", in R. Gleason, M. Laubhan and N. Euliss (2008), "Ecosystem services derived from wetland conservation practices in The United States Prairie Pothole Region with an emphasis on the United States Department of Agriculture Conservation Reserve and Wetlands Reserve Programs", United States Geological Survey.

Laubhan M., K. Kermes and R. Gleason (2008), "Proposed approach to assess potential wildlife habitat suitability on program lands", in R. Gleason, M. Laubhan and N. Euliss (2008), "Ecosystem services derived from wetland conservation practices in The United States Prairie Pothole Region with an emphasis on the United States Department of Agriculture Conservation Reserve and Wetlands Reserve Programs", United States Geological Survey.

Lubowski R., A. Plantinga and R. Stavins (2003), "Determinants of land-use change in the United States, 1982-1997", Discussion Paper 03-47, Resources for the Future, Washington DC.

OECD (2008a), *OECD Environmental Outlook to 2030*, OECD, Paris.

Osborn T. (1997), "New CRP Criteria Enhance Environmental Gains", *Agricultural Outlook*, Vol. 245.

Rabalais N., R. Turner and W. Wiseman (1997), "Hypoxia in the Northern Gulf of Mexico: Past, Present and Future", in *Environmental Protection Agency, Proceedings of the First Gulf of Mexico Hypoxia Management Conference*, Dec. 5–6, 1995, Gulf of Mexico Program Office.

Reynolds R., T. Shaffer, R. Renner, W. Newton and B. Batt (2004), *Impact of the Conservation Reserve Program on duck recruitment in the U.S. Prairie Pothole Region*, United States Fish and Wildlife Service.

Ribaudo M. (1986), "Consideration of off-site impacts in targeting soil conservation programs", *Land Economics*, Vol. 62.

Ribaudo M., D. Colacicco, L. Langner, S. Piper and G. Schaible (1990), "Natural resources and natural resource users benefit from the conservation reserve program", *Agricultural Economic Report*, Vol. 627, U.S. Department of Agriculture, Economic Research Service.

Roberts M. and S. Bucholtz (2005), "Slippage in the Conservation Reserve Program or spurious correlation: a comment", *American Journal of Agricultural Economics*, Vol. 87.

Samson F. and F. Knopf (1994), "Prairie conservation in North America", *BioScience*, Vol. 44.

Soil and Water Conservation Society and the Environmental Defence Fund (2008), "Conservation Reserve Program: Program Assessment", A report from the Soil and Water Conservation Society and the Environmental Defence Fund.

Sullivan P., D. Hellerstein, L. Hansen, R. Johansson, S. Koenig, R. Lubowski, R. McBride, D. McGranahan, M. Roberts, S. Vogel and S. Bucholtz (2004), "The Conservation Reserve Program: economic implications for rural America", *Agricultural Economic Report*, Vol. 834, USDA Economic Research Service.

Tangen B. and R. Gleason (2008), "Reduction of sedimentation and nutrient loading", in R. Gleason, M. Laubhan and N. Euliss (2008), "Ecosystem services derived from wetland conservation practices in The United States Prairie Pothole Region with an emphasis on the U.S. Department of Agriculture Conservation Reserve and Wetlands Reserve Programs", United States Geological Survey.

USDA (United States Department for Agriculture) (1997), "Agricultural Resources and Environmental Indicators, 1996–97", *Agricultural Handbook,* No. 712. USDA.

USDA (2002), 2002 USDA Agricultural Census, USDA National Agricultural Statistics Service.

USDA (2006), Fact Sheet, Conservation Reserve Program General Sign-up 33, Environmental Benefit Index, USDA, FSA, Washington.

US EPA (US Environmental Protection Agency) (2006), *Wetlands: Protection property and life from flooding*, Washington DC.

US EPA (2008), *Prairie Potholes*, US EPA, www.epa.gov

Wu J. (2000), "Slippage effects of the conservation reserve program", *American Journal of Agricultural Economics*, Vol. 82, No. 2.

Paying for Biodiversity: Enhancing the Cost-Effectiveness of Payments for Ecosystem Services

Chapter 7

Australia: The Tasmanian Forest Conservation Fund

Jim Binney[1] and Charlie Zammit[2]

This chapter presents the Tasmanian Forest Conservation Fund, a PES programme that aims to protect old growth forest on private land. Design elements, such as the use of a Conservation Value Index to identify areas of forest with high benefits and high threat of loss, and the use of inverse auctions to reduce the costs of obtaining these benefits are discussed. Finally, the chapter discusses the lessons learned and how these are being applied in the Environmental Stewardship Programme.

1. Marsden Jacob Associates, Brisbane, Queensland.

2. Department of the Environment, Water, Heritage and the Arts, Canberra, ACT.

7.1 Introduction

Australia has a long history of environmental debate over forest use, including over the conservation of old growth forests (Dargavel, 1995). In 1992 a national policy framework, *The National Forest Policy Statement,* was agreed between the Australian Commonwealth and all state and territory governments. Thereafter a series of twenty-year Regional Forest Agreements were progressively established by the Commonwealth and specific state governments between 1997 and 2001 to manage the long term protection and sustainable use of the nation's tall forest estate.[1]

A Tasmanian Regional Forest Agreement was finalised by the Australian Commonwealth and Tasmanian Governments in 1997. Following a review in 2002, a supplementary Agreement was put in place in 2005. Under the supplementary Agreement an additional 135 450 hectares of forest was identified for protection of which the majority was sourced from public forest land. However, the Agreement also identified the protection of up to 45 600 ha of forest on private land to be achieved through voluntary market-based measures. The Forest Conservation Fund was created to meet this policy objective.

The Forest Conservation Fund

The Forest Conservation Fund (the Fund) comprised a suite of market-based approaches to secure the protection and management of high conservation value forests on private land in Tasmania. The Fund included:

- PES mechanisms: inverse auction, differentiated take it or leave it offers, and direct negotiation approaches; and
- the establishment of a revolving fund for the purchase, protection and resale of high conservation properties in the existing property market.

The focus of this case study is the Fund's PES mechanisms. The total budget available for the Fund was approximately AUD 50 million. The primary target for the Fund was to protect up to 45 600 hectares of forested private land, targeting old growth forest and forest communities known to be under-reserved in the public protected area system. Accordingly, the Fund specifically aimed to protect:

- a minimum of 25 000 hectares of old growth forest; and
- up to 2 400 hectares of forest to protect the karst values in the Mole Creek area.

The case for market failure in nature conservation and the protection of native forests has been extensively made, specifically relating to the 'public good' aspects of native vegetation (see for example Productivity Commission, 2004). Other key drivers for government intervention and the decision to use a competitive inverse auction approach included:

- insufficient market incentives for the protection of socially optimal levels of environmentally valuable forest assets on private land;
- the heterogeneous nature of environmental values attached to different areas of forest;
- the heterogeneous nature of the opportunity costs (forestry production foregone and management costs);
- limitations on the budget available to achieve the conservation targets; and
- problems of information asymmetry, particularly hidden information that may result in adverse selection problems.

7.2 Key design elements of the FCF

The success or failure of the Fund is highly reliant on the ability to create and run an efficient market for the protection of forest on private land in Tasmania. The Fund was designed through a policy implementation process supported by rigorous analysis by a number of experts with significant knowledge and skills in ecology, forestry practices, geographical information systems and ecosystem mapping, economics and market based instruments A number of possible assessment and market approaches were considered before a decision on the final design of the Fund was established (AMAP, 2006). Key elements of the Fund are shown in Table 7.1.[2]

Design of the Fund on-ground implementation process

The process for the implementation of the Fund was also carefully designed, drawing on the knowledge and experience of national experts with significant experience in PES schemes. Key aspects of the Fund implementation process are shown in Figure 7.1.

Table 7.1. **Key elements of the FCF**

Issue	Key design element
Mechanisms	Inverse auction. Several rounds were conducted. Rounds 1a to 1c from the initial pool of participants and Round 2 from a later pool of participants. Following round 1c of inverse auction, differentiated take-it-or-leave-it offers were made to landholders. Direct approaches through a third party service provider.
Price	For inverse auction: landholder paid their own winning offer prices. For differentiated take-it-or-leave-it, prices based on modelled values of equivalent successful bids from inverse auction rounds 1a to 1c. For direct approach, price was that agreed by both parties.
Bids	Sealed bids.
Rounds	Multiple auctions conducted until available budget was exhausted and/or targets achieved.
Assessment of conservation values	Specific assessment metric created – the Conservation Values Index (CVI).
Bid selection	Based on unit cost of conservation benefits from individual bids (AUD/CVI).
Securing property rights	Two mechanisms used: a covenant attached to the land title that binds current and future owners; and a management agreement that outlines agreed management actions to enhance forest condition and extent.
Selection cut-off	No formal price cap used, but cut-off for each round established at natural point of inflection in aggregate cost curve from that round.
Decision-making	Fund Assessment Panel, supported by technical experts considers all bids and recommends to Minister for the Environment for funding approval.
Payments	Ex-ante (20% on signing agreement and 80% on registration of covenant).
Ongoing monitoring, reporting and evaluation	Requirements on landholder to report on management actions. Ongoing monitoring and evaluation undertaken by Tasmanian Government.
On-ground delivery of Fund	A third party delivery model was adopted to ensure local presence and on-ground capacity in Tasmania.

Source: OECD, 2010.

Figure 7.1. **Simplified representation of FCF implementation process**

Participant engagement

- Public information and awareness programme undertaken.
- Expression of interest in participation in Fund lodged by landholder.
- Fund information kit provided to landholder.

Proposal development

- Field Conservation Advisor visits landholder's property. Conservation values of eligible forests assessed using standardised ecological field protocols and conservation management options are discussed with land owner. Note: conservation advisers did not discuss financial matters or bid prices with landowners.
- Landholder develops a bid proposal for identified areas to be protected, management actions to retain and enhance condition and the proposal price.
- Landholder submits sealed bid to tender process.

Selection process

- Proposals are assessed using the CVI metric based on information from the field assessment, mapped forest asset data, reservation status, range of management options proposed and length of covenant offered.
- Proposals for each round are then ranked based on value for money (AUD/CVI) and reviewed by the Fund Assessment Panel to establish a cut-off level for successful proposals for each round (see Figure 7.2). Recommendations for funding are then made to the Commonwealth Minister for the Environment.

Contract implementation

- Financial offers made to successful proposals. If accepted, conservation agreement and covenant finalised.
- Once the contract between the government and landowner is signed, initial payment (20%) made.
- The covenant is registered on land title and final payment (80%) made.
- Landholder undertakes ongoing implementation of land management as per agreements and ongoing support and compliance management is undertaken by Tasmanian Government.

Source: OECD, 2010.

Attracting participants and creating competition

Prior to the implementation of the Fund, only limited knowledge was available on the number of potential participants, their willingness/ability to develop bids with high environmental values, and the potential degree of price competition.

Information materials for participants were specifically designed to both encourage participation and assist in the development of quality bids. This included media (print and radio); information packs; and public information sessions to explain the Fund target forest communities, mechanism and processes. The fact that the Fund was a suite of market mechanisms (inverse auctions, direct negotiations, revolving fund etc.) was emphasised to ensure landholders were aware of their options for participation.

The competitive elements of the Fund were continually emphasised, particularly that available funding was limited, competition for funding would be high, and that bids would be selected based on value for money.

When implemented, the Fund was initially overwhelmed with in excess of 420 expressions of interest and approximately 240 requests for site assessments. This far exceeded the capacity of the on-ground delivery team, and was a key reason for establishing multiple rounds to make the task manageable. In hindsight, as part of the design process, it would have been prudent to:

- undertake more detailed market assessments to estimate likely participation rates; and
- establish processes to manage the potential for over/under-subscription of the Fund.

A robust metric – the Conservation Value Index (CVI)

The metric developed for the Fund – the Conservation Value Index – was based on the objectives outlined in the Fund Strategic Plan (Commonwealth of Australia, 2006) with significant input and research by nationally recognised experts. The CVI was developed to assess three aspects of a landholder's proposal:

- *significance* of the proposal in contributing to the conservation objectives of the Fund;
- *conservation management* provided by the proposal in relation to current conditions and risks that *would not* have been undertaken in the absence of the FCF; and

- *security* of the proposal measured as the covenant length offered (12, 24, 48 years and in-perpetuity).

The CVI includes key criteria to assess each proposal against the objectives of the Fund, particularly:

- *forest priority Score* assesses the relative preferences for different forest types, prioritised against their conservation status;
- *structural condition* details the structural form of the forest, derived from assessments of Regional Forest Agreement Forest Resource Types;
- *current condition* of the proposal areas based on benchmarked forest conditions;
- *regional threat index* assesses the threat to the proposed forest area from surrounding land uses and conditions;
- *reservation* considers the current level of protection for each specific forest type using the established regional forest agreement reservation target system (Commonwealth of Australia, 1997);
- *maintenance* determines a value to represent maintenance of current forest condition;
- *improvement* considers the voluntary management actions and the impacts they are likely to have on improving the condition of the proposal site; and
- *security* measures the duration of security offered by the proposal to ensure conservation values are achieved for either a fixed term or a perpetual covenant.

Models were developed that calculated the CVI for each proposal to enable ranking of proposals based on a value for money criteria (AUD/CVI). Weightings in the CVI were based on known or modelled relationships between key attributes of forest conservation and also the consensus opinion of national experts (AMAP, 2006, Eigenraam *et al.,* 2007).

The CVI is theoretically robust, practical, repeatable, transparent, and pragmatic given the data, knowledge, and programme constraints. Given time, information, and budget constraints, it is unlikely a materially better metric could have been developed at the time. However, during the implementation of the Fund, potential enhancements to the eligibility criteria and CVI were identified, particularly where the assessment process could be simplified or modified without losing any functionality or ability to differentiate between proposals. For example, to increase efficiency and reduce risk, the eligibility criteria for funding proposals was tightened over time - to increase the minimum area to be covenanted and set minimum security at 24 years. These changes to eligibility criteria were addressed through calculated CVI scores.

The CVI and field assessment

The utility of the CVI to differentiate between proposals is highly reliant on the data collected through the field assessments. A number of actions were undertaken to ensure the quality and appropriateness of assessments including:

- employing field officers (Conservation Advisors) that had appropriate formal qualifications and survey experience (*e.g.* forest ecology, forest and/or conservation management);
- formal training was provided for all Conservation Advisors in the on-ground application process and development of data for the CVI;
- the development of a specific field assessment manual to assist with on-ground assessments; and
- a process of quality assurance to ensure consistency in the assessment between Conservation Advisors and the comparability of all proposals received.

These actions reduced the risk of poor data quality impacting on the assessment process.

The treatment of transaction and administrative costs in value for money assessments

Transaction and administrative costs for PES schemes can sometimes be significant, particularly where detailed field assessments, specific legal documentation (*e.g.* covenants) and ongoing monitoring are required. Most costs incurred in attracting and assessing proposals cannot be easily avoided irrespective of the success/failure of the proposal. However, future management costs, including ongoing monitoring, evaluation and compliance, can also be significant, and are often fixed in nature, irrespective of the conservation values of the proposal.

As part of the mid-term review of the Fund, the potential impact of future transaction and administrative costs was identified as a potential area where the life-cycle efficiency of the Fund could be impacted (Marsden

Jacob Associates, 2010). Sensitivity analysis of proposals from round 1a including estimated future administrative costs was undertaken[3] and the rankings were compared to the actual rankings used. The rankings of some proposals did change when future administrative costs were included, although no accept/reject decisions would have changed.

While the sensitivity analysis found that the inclusion of future administrative and transaction costs was not warranted for the Fund, this issue may warrant consideration in the design of future PES schemes. In particular, this is likely to be relevant for schemes aiming to invest over time in significant ecological restoration of high conservation value assets.

The CVI and broader area based targets

As noted, the forest conservation targets identified as part of the Tasmanian Supplementary Regional Forest Agreement are area based. However, the assessment, prioritisation and selection of Fund bids is based on a cost-effectiveness metric (*i.e.* AUD/CVI). Area based targets, while easier to identify, can be an inferior indicator of conservation value as they only consider the extent of forest protection achieved. The CVI is a superior measurement as it considers forest extent and condition, and in particular both current condition and future condition when management actions are in place.

The potential inconsistency between area-based conservation planning targets and the selection of proposals based on cost-effective metrics highlights the need to educate decision makers and the community of the relative merits of using metrics to drive public funds in conservation.

Selection of proposals

The design of the selection process involved a governance framework overseen by the Fund Steering Committee, comprising senior officials from the Commonwealth and Tasmanian Governments and supported by an external probity advisor. The probity adviser was responsible for ensuring fair and transparent programme implementation and was available for advising on any disputes between landholders and programme managers and service provider contractors.

The selection of proposals for the Fund involved a number of steps, specifically:

- Individual proposals were assessed using the CVI based on information from the on-ground assessment, mapped forestry asset data, reserves status, and length of covenant offered etc.

- Proposals for each round were then ranked based on cost effectiveness (AUD/CVI) and reviewed by the Fund Assessment Panel.
- A cut-off level for successful proposals for each round was established, based on the point in the aggregate supply curve for that round where the cost of bids (AUD/CVI) increased rapidly (see Figure 7.2).[4] The Fund Assessment Panel also reviewed all proposals to ensure proposals were consistent with the Fund's objectives and principles.
- Recommendations for funding were then made by the Fund Steering Committee to the Commonwealth Minister for the Environment.

Contractual arrangements and ongoing monitoring and evaluation

The Fund was underpinned by two key contractual agreements:

- a covenant attached to the land title deed held by the land owner provided the primary security to protect and manage forest assets; and
- a financial agreement for payments from the Commonwealth Government to the landholder.

Covenants are documents that govern land use and may impose conditions upon the management of a specific parcel of land. They are legally binding on current and future landholders and are registered on the land title under the Tasmanian *Nature Conservation Act 2002.* A specific design element of the Fund was to offer a choice of covenant length (12 years, 24 years, 48 years and in-perpetuity). The rationale for offering multiple lengths was to enhance participation in the Fund. Later in the implementation of the Fund, the 12 year option was dropped as it provided limited conservation benefit and had proved unpopular.

The financial agreement included two ex ante payments: 20% on signing a letter of acceptance and contract; and the further 80% once the covenant was registered on the land title.

Semi-structured interviews with participants, undertaken as part of the Fund's mid-term review, indicate that the ex ante payments were popular, but the payment stream was misaligned with the actual costs faced by some landholders. This has the potential to exacerbate compliance risks for the Fund. Recognising this risk, the Australian Government is now utilising contacts that include both ex ante payments (representing capitalised production values foregone) and ex post payments for management actions undertaken.

The design of the Fund also included ongoing management and support services to landholders provided by the Tasmanian Government. The Tasmanian Government is also responsible for the ongoing compliance management, monitoring and reporting of forested lands covenanted under the Fund. These services were developed under a separate contact between the Commonwealth Government and the Tasmanian Governments.

7.3 Effectiveness and efficiency of the FCF

Key achievements

Over the life of the Fund landholders requested approximately 420 information kits, leading to 240 site assessments. For the inverse auction rounds, a total of 183 full bids were received, of which 95 (52%) were successful. The variance of bid values from the inverse auction also indicates significant price competition amongst participants. Of the 88 unsuccessful applications in the inverse auction rounds 1a to 1c, 26 landholders subsequently accepted differentiated take-it-or-leave-it offers. A further eight direct offers were negotiated.

Areas protected

The total areas secured by the Fund are summarised in Table 7.2. The Fund secured a significant area of high quality forest on private land, totalling almost 29 000 hectares from a target of up to 45 600 (63%). With a stated target of securing 25 000 hectares of old growth forest, the Fund secured almost half of this (11 000 ha).

Table 7.2. **Area secured by the FCF**

Forest type	Target (ha)	Secured (ha)	% of target	Outstanding (ha)
Total	(up to) 45 600	28 900	63	16 700
Old growth	25 000	11 000	44	14 000

Source: OECD, 2010.

While a significant achievement in itself, to completely satisfy all of its targets, the Fund would need to secure an additional 16 700 hectares, predominantly old growth forest. A longer term revolving fund[5] has been established to progressively address this shortfall as market opportunities arise and Fund criteria are met.

Duration of protection secured

Landholders were able to nominate the length of covenant they were prepared to enter into (12 years, 24 years, 48 years, in-perpetuity).[6] The Fund CVI assigned a greater weight to longer covenants over those of shorter periods. Areas secured and the duration of protection are shown in Table 7.3. The majority of area secured was in-perpetuity (over 24 000 hectares or 80% of the total). Covenants made for 48 years totalled only 2% of area, suggesting they were a less valuable option. The 12 and 24 year covenants accounted for the remaining 13% of areas.

Semi-structured interviews undertaken with a sample of landholders participating in the Fund indicate the major reason for choosing a shorter length covenant was to ensure options for future generations of landholders were not extinguished.

Table 7.3. **Duration of protection secured by the FCF**

Duration	Total area (Ha)	AUD/Ha	AUD/CVI
Perpetuity	24 225	AUD 1 775	0.28
48 years	682	AUD 1 570	0.32
24 years	3 614	AUD 604	0.40
12 years	295	AUD 331	0.74

Source: OECD, 2010.

Overall, the cost-effectiveness of bids was higher for longer term contracts. This is largely driven by the higher weightings placed on longer term contracts within the CVI more than offsetting the increase in bid prices offered. The effectiveness and efficiency of the Fund was impacted by a number of factors including:

- the effectiveness of the on-ground delivery;
- the relative efficiency of the PES mechanisms used (*i.e.* inverse auction and differentiated take-it-or-leave-it approaches);
- the transaction and administration costs in running the Fund; and

- the way in which participants engaged in the Fund and how that impacts on the cost of their proposals.

On-ground delivery

The Commonwealth Government does not have capacity in Tasmania to undertake the on-ground delivery of the Fund. In addition, the relatively short timeframe for the implementation of the Fund (less than three years) and variability in skill requirements over the life of the programme precluded quickly establishing an experienced in-house team based in Tasmania. Given this, a decision was made to have the on-ground delivery of the Fund delivered by third party organisations.

Detailed specifications of the requirements for the on-ground delivery were developed and an open tender was used to select and procure the services. Two organisations were awarded contracts:

- A consortium led by a multinational services firm. The consortium included skills in ecology, GIS, communications and business and programme administration. This consortium was responsible for the delivery of the inverse auction rounds and the take-it-or-leave-it offers of the PES scheme.
- An environmental non-government organisation to manage the direct approach component of the PES (run concurrently with Round 2 of the inverse auction).

Each of the third party service providers worked closely with relevant officials of the Commonwealth and Tasmanian Governments to ensure their contractual obligations were performed and the operational objectives of the Fund were achieved.

Effectiveness of third party on-ground delivery

Both parties undertaking the on-ground delivery attracted significant numbers of quality proposals into their respective Fund programmes. The third party delivery model had some distinct advantages, particularly the ability to utilise existing corporate infrastructure, networks and local technical knowledge. Independent evaluation of the Fund found the third-party delivery model was generally effective (Marsden Jacob Associates, 2010). However, a number of operational problems did arise which required resolution during the Fund delivery phase. Key lessons that emerged from the use of a third-party delivery organisation included:

- The need for more accurate specification of the roles, responsibilities and requirements of third party delivery organisations. At times problems were faced where the respective responsibilities were ambiguous, creating delays and bottlenecks in on-ground delivery.

- Given the nature of the objectives of the Fund and the target participants (*i.e.* primarily foresters and farmers) it is vital that the on-ground delivery agent has an understanding of the target participants and their industry to maintain credibility and develop the market. Semi-structured interviews with participants raised particular problems with a lack of knowledge of farming, forestry and conservation within critical parts of the consortium delivery organisation. This may have had a detrimental impact on the conversion of expressions of interest into actual proposals.
- Contingencies need to be in place to mange under/oversubscription of programmes and the variability in workloads. Where this is not done, delays in processing and assessing proposals can occur and the credibility of the PES programme can be impacted.
- The need to maintain consistency in the quality of work undertaken is vital. This is particularly the case with direct interaction with participants and technical field work. Where resources of a sufficient quality are limited, an assessment of the tradeoffs between extending programme timelines (*e.g.* running multiple rounds) versus the quality of work will need to be considered.

Inverse auction outcomes

The inverse auction efficiently secured conservation outcomes on private land. Key statistics are outlined in Table 7.4. Major points to note include:

- There was significant variance in bid prices (measured by AUD/CVI) in all rounds, reinforcing the decision to use a inverse auction approach to help reveal true opportunity costs. This is consistent with the outcomes of the semi-structured interviews with landholders that indicated the heterogeneous nature of the opportunity costs.
- There was a general increase in average bid prices between rounds. This is partially explained by price learning effects in the market as the Fund progressed. However, it is also partially explained by a number of landholders participating in Rounds

1a to 1c that already had a history of conservation on private lands and participation in prior incentive mechanisms.

- The interviews revealed a number of approaches were used to price bids. These were: bids based on the full commercial opportunity cost (particularly from larger land holdings); bids that only reflected management costs (particularly for smaller 'lifestyle' holdings); and bids that reflected attempts to inverse engineer a maximum acceptable price (based on CVI scores (provided to landholders) and hearsay regarding prices paid for winning bids in previous rounds).

Table 7.4. **FCF inverse auction - key statistics**

Round	Area (ha)	CVI (total)	AUD/ha	AUD/CVI	Bids wins (and total)	AUD/CVI range
1a	3 921	17 750 000	925	0.20	24 (36)	AUD 0.07-A UD 0.81
1b	3 192	14 647 000	1 168	0.25	26 (58)	AUD 0.04-A UD 0.49
1c	1 916	6 465 000	1 270	0.38	16 (49)	AUD 0.16-A UD 1.14
2	4 750	18 272 000	1 683	0.44	29 (40)	AUD 0.23-A UD 0.71
Total	13 779	57 136 000			95 (183)	
Average			1 290	0.31		

Source: OECD, 2010.

Figure 7.2 shows the cumulative cost curve for each round of the inverse auction, ranked by most cost-effective bid to least for each round. The horizontal solid lines are the cut-off point for each round.

Individual curves transit vertically at the localised area where cost effectiveness falls away for each round. As the graphs show, each round cuts off at broadly similar points of cost effectiveness, suggesting there was limited scope for efficiency gains from different cut-off points between rounds. The variance in bid prices across all rounds indicates that the competitive nature of the Fund was maintained across all rounds. By the end of Round 2, the backlog of bids had essentially been cleared. If further time was available, the only means to maintain or improve cost effectiveness would have been to completely re-open the bidding process and introduce new market participants.

Figure 7.2. **FCF supply curves for conservation (inverse auction rounds)**

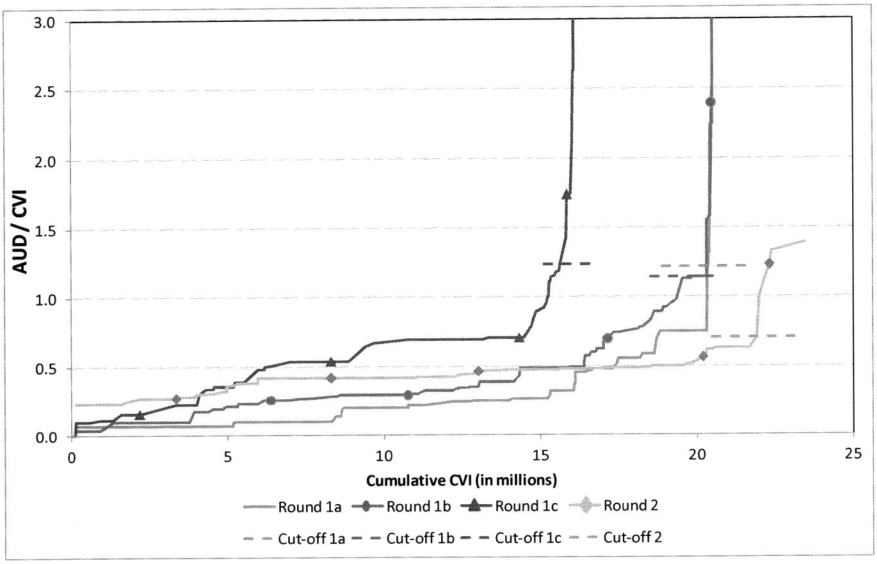

Source: OECD, 2010.

Comparisons of inverse auction vs. other approaches

At the conclusion of Round 1c of the inverse auction, it was clear that the Fund was running behind schedule in meeting its targets. To expedite the program, a decision was made by the Fund Steering Committee to also utilise two other approaches in parallel with a further round of the inverse auction. These were:

- Differentiated take-it-or-leave-it offers to unsuccessful bids from rounds 1a to 1c. These offers were based on the modelled AUD/CVI from successful bids under rounds 1a to 1c of the inverse auction. Participants had the choice to accept the offer (guaranteed success, but potentially inadequate revenue stream); resubmit a different a bid in the Round 2 inverse auction (uncertain outcome); or reject all offers.

- Direct approach offers were made through a third party service provider to a number of larger landholders with known high conservation value forest assets. Offers were again based on the modelled AUD/CVI from rounds 1a to 1c of the inverse auction.

Prices offered for the differentiated take-it-or-leave-it offers were estimated through the application of a non-linear regression model of AUD/ha and CVI/ha for all successful bids from rounds 1a to 1c of the inverse auction. This model could be reliably applied to any bid where forest area and CVI score was available.

Because the various Fund PES components were run in a relatively similar area within a narrow time frame and all used the CVI to measure conservation benefits, analysis of the data can provide some important insights into the efficiency of the market approach in eliciting cost-effective bids.

Table 7.5. **FCF inverse auction vs. other approaches**

Approach	Area (ha)	CVI (total)	AUD /ha	AUD/CVI	Bid wins	AUD/CVI range
Inverse auction	13 779	57 136 000	1 290	0.31	95	AUD 0.07-AUD 1.14
Direct approach	5 657	43 132 000	1 700	0.22	8	AUD 0.21-AUD 0.24
Take-it-or-leave-it	2 996	18 106 000	1 418	0.23	26	AUD 0.19-AUD 0.34

Source: OECD, 2010.

Table 7.5 summarises the key statistics for the inverse auction and the other approaches (differentiated take-it-or-leave-it and direct approach). Unsurprisingly, the direct and differentiated take-it-or-leave-it approaches had a narrower range of costs than the inverse auction due to the limitation imposed on price variation. They also had a lower average cost per CVI than the inverse auction approach. However, it is important to note that the direct and differentiated take-it-or-leave-it approaches would not have been possible in the absence of the inverse auction rounds (1a to 1c) as the opportunity costs were essentially unknown prior to the commencement of the Fund. In effect, rounds 1a to 1c were needed to create a market and for 'price discovery'. The direct approach is also characterised by large areas, which were intentionally targeted, a higher price per hectare but a low price per CVI. This was partly driven by the requirement that all direct approach offers include an in-perpetuity covenant. This is also a feature of the differentiated take-it-or-leave-it approach.

While the data indicates that the direct approach and differentiated take-it-or-leave-it approaches had lower costs (in AUD/CVI), they may have also created some unintended outcomes, specifically:

- A floor price (minimum price) that was potentially unrelated to opportunity costs for some participants. Semi-structured interviews with landholders indicated that a 'market clearing' price had now essentially been created and that future programmes may struggle to elicit bids below that price.
- Because the differentiated take-it-or-leave-it offers were available at the same time as Round 2 of the inverse auction, they may have also moderated the potential for rent seeking by landholders intending to participate in the inverse auction. This may partially explain the narrowing of the variation in bids in the inverse auction for Round 2.
- For the differentiated take-it-or-leave-it offers made to unsuccessful participants in Rounds 1a to 1c, there is a potential compliance risk where payments made are actually lower than efficient opportunity costs and difficulties may arise in meeting long-term contractual obligations.

A comparison of the conservation benefits achieved (in CVI) and relative costs of the different approaches (average AUD/CVI) are shown in Figure 7.3. Key points to note include:

- Round 1a (inverse auction) secured 15% of the conservation gains at relatively low cost, partially due to the number of 'early adopters' with a strong conservation ethic participating.
- The Round 2 direct and differentiated take-it-or-leave-it approaches offers secured approximately 52% of the conservation gains. These approaches were run after Rounds 1a to 1c and reflect the competitive market prices emerging from those early inverse auctions.

Figure 7.3. **Conservation gains and relative costs between FCF inverse auctions vs. other approaches**

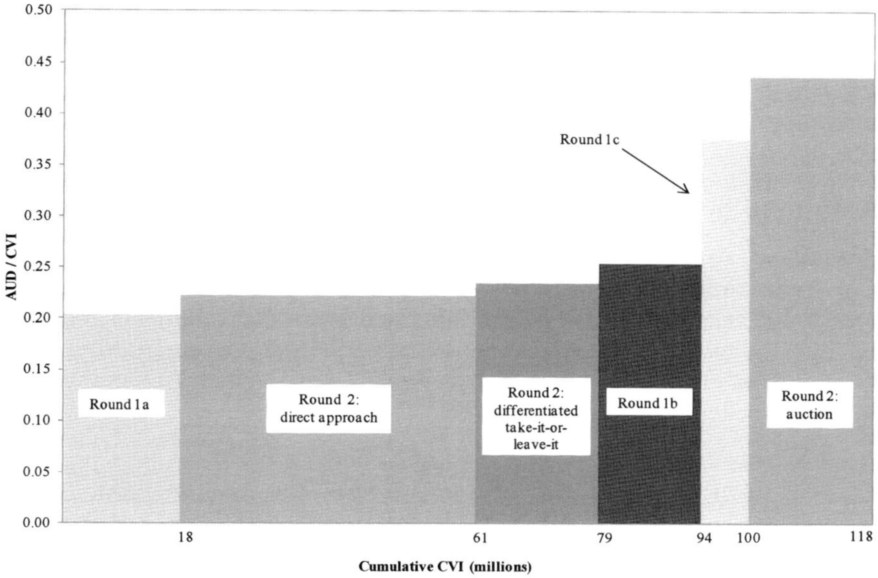

Source: OECD, 2010.

Transaction, management and administration costs

One common criticism of PES mechanisms is based on the perception that they involve higher transaction and administrative costs than more traditional funding models, for example through devolved grants. These additional costs generally relate to the development and operation of more sophisticated market approaches and metrics. However, market approaches such as PES schemes have the potential to provide more cost-effective outcomes, where the additional management and administrative costs are less than the value of the gains in conservation outcomes.

Costs for Fund management and administration

Some management and administrative costs were largely fixed, while some were variable depending, for example, on the number of proposals and the property size. A bottom up accounting model was developed to estimate the Fund management and administration costs covered by the Australian Government (Marsden Jacob Associates, 2010). Costs have been attributed over completed transactions. Table 7.6 shows a breakdown of relevant management and administration costs. Key findings include:

- The cost of designing and administering the PES components of the Fund was 10.5% of the total relevant programme budget.
- Approximately 46% of the total programme administration costs were for general programme administration, proposal assessment and communications. These costs would generally be incurred irrespective of the market instrument used. To a certain extent these costs are higher than could be expected, reflecting the fully commercial nature of the major on-ground delivery organisation and the fact that administrative systems had to be established and operated specifically for the Fund.
- Further analysis of management and administrative costs for each sub-element of the Fund and each inverse auction round shows costs varied significantly due to the complexity of administrative tasks, and the ratio of assessments to eventual accepted offers.
- As would be expected, the detailed field visits were also a major cost driver (21.7% of total management and administrative costs). However, these costs could not been materially reduced as this function provided critical inputs to proposal developments and CVI calculations.

Table 7.6. **FCF management and administration costs**

Cost category	% of management and administration costs	% of Fund PES component budget
Technical design & advice (including CVI)	5.2	0.5
Legal expenses including covenants	11.5	1.2
GIS inputs	10.5	1.1
On-ground site assessments	21.7	2.3
Independent probity inputs	2.4	0.2
Administration, proposal assessments & communication	46.1	4.8
Independent evaluations	2.7	0.3
Total	100.0	10.5

Source: OECD, 2010.

Transaction costs faced by participants

Landholders participating on the Fund also faced their own transaction costs. While quantitative data is not available on these costs, semi-structured interviews (Marsden Jacob Associates, 2010) did gain some insight into those costs. Key findings from that analysis include:

- Landholder transaction costs were highly variable, depending on specific property circumstances (*e.g.* whether they had comprehensive information and valuations of their forest assets), the level of consultation undertaken with family members (*e.g.* discussing property succession options with their children), and the degree to which professional advice was sought (*e.g.* tax advice, property valuation advice).
- Developing the content of proposals generally took between a few hours and a few days of actual time inputs. Often these time requirements were increased as participants sought additional information regarding rights and obligations under the programme.

While all successful and unsuccessful participants faced transaction costs, interviews revealed that even the unsuccessful participants gained some benefit from the program, particularly a better understanding of the extent and condition of the forest assets on their property and a better understanding of best management practices to maintain or enhance forest condition.

Efficiency gains from the PES approach

There are two major potential efficiency gains from the PES approaches used in the Fund. First, the additional conservation gains made from using the inverse auction approach. Second, the additional gains from using the environmental metric, the CVI.

Additional conservation gains from using the inverse auction approach

The major policy innovation in the Fund was the use of an inverse auction to create a competitive market and to ensure value for money. It is possible to estimate the efficiency gains from the inverse auction by comparing successful bids using the auction rounds compared to a less sophisticated approach to incentive design, for example, awarding contracts in the order in which proposals with appropriate forest types are received. Table 7.7 shows the total value of CVI units purchased through the inverse auction rounds of the CVI, compared to the CVI units that would have been purchased if proposals had been funded based on the order in which they were received.[7] It demonstrates that the gains from using the inverse auction approach can be very significant; in this case, in excess of 52%.

Table 7.7. **Potential conservation gains from the FCF auction approaches used**

Conservation Outcomes	AUD millions
CVI units purchased using inverse auctions (millions)	90.8
CVI units purchased where selection are based on order of proposals received (millions)	59.6
Increase in CVI units from use of inverse auction (millions)	31.2
Increase in CVI units (%)	*52.3%*

Source: OECD, 2010.

Return on investment in CVI-based selections

One of the criticisms of sophisticated PES schemes like the Fund is the significant up-front investment often required to design metrics and the additional GIS inputs associated with applying the metric. It can be argued that all of the other management and administration costs would be the same for a PES program, irrespective of the metric used. Therefore it is possible to isolate the efficiency gain from using a more sophisticated metric, where:

- benefits are valued based on differences in conservation gains between selections using a complex metric (in this case AUD/CVI) and a simple selection process (say AUD /ha); and
- costs are the incremental management and administration costs attributable to the design and application of the metric to underpin the selection process.

Using actual proposal data from the Fund, selections of the most cost-effective proposals were made using AUD/CVI (a complex metric) and AUD /ha (a simple metric) assuming a hypothetical AUD 20 million programme budget.[8] The value of additional CVI units achieved using the AUD/CVI metric are estimated based on the average AUD/CVI from all successful bids. Results of this hypothetical analysis are shown in Table 7.8.

Using the AUD/CVI metric, an additional 18.6% in conservation outcomes are achieved. The additional conservation gains are valued at approximately AUD 3.3 million, while the cost of achieving those benefits is only AUD 0.5 million. The ratio of benefits to costs from investing in the CVI is 6.9:1.

Table 7.8. **Analysis of CVI return on investment in the FCF (hypothetical AUD 20 million programme)**

	AUD millions
Conservation outcomes	
CVI units purchased using AUD/CVI selection (millions)	66.3
CVI units purchased using AUD/ha selection (millions)	55.9
Increase in CVI units (millions)	10.4
Increase in CVI units (%)	*18.6%*
Economic benefits and costs	**USD (millions)**
Estimated value of additional conservation outcomes (millions)	AUD 3.3
Estimated incremental cost of establishing and using CVI (millions)	AUD 0.5
Net benefit from CVI based assessments (millions)	AUD 2.8
Benefit cost ratio	*6.9:1*

Source: OECD, 2010.

Landholder engagement

As outlined in Section 7.2, a key element of the design of the Fund was the extensive effort undertaken to design a PES scheme that effectively engaged landholders and elicited value for money forest conservation outcomes. Semi-structured interviews (Marsden Jacob Associates, 2010, Ipsos, 2009) investigated a number of issues relating to engaging landholders, the design and implementation of the Fund, and the impact it had on proposal prices and covenant lengths. A number of key findings emerged from the analysis of interview results.

Information provision: content and approach

As noted, the Fund provided significant public information to inform the market. The level and structure of this information can have an impact on participation levels and proposal prices.

Semi-structured interviews indicated that the level of satisfaction with printed information available (Commonwealth of Australia, 2007) was generally high, but that the language could be simplified and more case studies provided. However, there were critical issues where information was not readily available, particularly the tax treatment of payments and the potential for capital loss implications. In addition, some landholders that sought professional advice found the advice expensive to obtain and in some cases ambiguous. This may have had an upward impact on proposal prices as opportunity costs were worked out on a pre-tax cost base, whereas the true financial cost to the landholder was often on a post-tax basis.

These issues raise the need to ensure a broad suite of fit-for-purpose information products are available for all critical issues that impact on participation and proposal prices.

In addition to the various levels of printed information, there were two key forms of verbal information available to participants; formal information sessions and direct contact with Conservation Advisors, primarily during property visits. Generally the interviews revealed that the information sessions could be significantly improved by providing more in-depth information, for example through an introductory session and an in-depth session, and ensuring presenters have significant industry knowledge and credibility. Field assessments and one-to-one contact with Conservation Advisors were generally very well received.

Establishing reasonable proposal prices

PES programmes will be most efficient where proposal prices are an accurate reflection of economic opportunity costs. While the competitive nature of the Fund discourages rent seeking behaviour, interviews revealed that participants often incorporated a contingency cost or uncertainty premium within their proposal prices. Key drivers of these contingency values included:

- commercial issues such as taxation treatment (mentioned previously) and impacts on property values and property rates;
- the 'fit' of obligations under the Fund with broader property management and landholder aspirations;
- a reluctance to commit their children to obligations under the Fund (particularly for 48 year and perpetual covenants);
- uncertainty regarding the costs of some management actions in the long-term (*e.g.* costs of replacing fences to exclude stock in 50 years); and
- limited capacity to systematically establish a proposal that effectively meets the requirements of the Fund (*e.g.* which management actions should be included) and represents the tradeoffs between commercial outcomes and delivering environmental services.

In addition to the upside price risks, where participants are unable to establish a reasonable proposal price, and bid below the true economic cost, they are more likely to become a compliance risk to the programme in the future.

The limited capacity of some participants to establish a reasonable proposal (both content and price) could have a significant impact of the efficiency of the PES scheme. Therefore, it would be prudent to undertake modest investments in enhancing participant capacity. For example, workshops to assist participants in resolving any uncertainty without perversely impacting on the competitive nature of the programme.

PES instrument used: inverse auction vs. other approaches

Tasmania has a history of utilising grants-based funding mechanisms and suasive programmes to encourage enhanced forest conservation on private land. The Fund was the first attempt to use a more sophisticated PES approach. Semi-structured interviews reveal mixed preferences towards the two approaches.

Many participants, particularly landholders on larger properties, preferred the ability to establish a price themselves under the inverse auction approach. The inverse auction approach overcame common shortfalls between private costs and funding available under other programmes with co-contribution ratios (*e.g.* 50% landholder and 50% government).

Conversely, many other participants struggled to establish a price and/or were opposed to the highly competitive nature of the inverse auction. These landholders held a strong preference for the differentiated take-it-or-leave-it approach. However, it should be noted that the introduction of take-it-or-leave-it offers in Round 2 of the Fund created dissatisfaction amongst some participants from the earlier rounds that had submitted successful proposals at a lower price than the those offers. It may have also perversely encouraged an upward shift in price expectations for some landholders who had opportunity costs below the take-it-or-leave-it rate being offered.

The different preferences of participants in the Fund reinforced the decision to establish a suite of PES and other market based approaches under the Fund, each with different attributes that would appeal to a wide mix of landholders.

7.4 Application of lessons in the Environmental Stewardship Programme

The Fund was the first major Australian Government market-based scheme to protect biodiversity. While a significant investment was made in the design process, the Fund was also subject to ongoing monitoring, evaluation and adjustment over the life of the programme.

In 2008 the Australian Government announced the Environmental Stewardship (ESP) programme as part of the national Caring for our Country environmental initiative of more than AUD 2 billion over five years.[9] The Environmental Stewardship programme continues the use of inverse auctions to protect high conservation value assets on private land. However, it diverges from the Forest Conservation Fund in several critical ways:

- Its scope is restricted to investments in matters of national environmental significance as defined under the Commonwealth Government's *Environmental Protection and Biodiversity Conservation Act 1999.* These include nationally endangered ecologically communities and species. The first rounds of the Stewardship Programme have targeted the nationally endangered box-gum woodlands of south-eastern Australia, and new rounds are targeting multiple ecological communities in other regions.
- Contracts agreed through the auction process provide annual payments to land managers for up to 15 years, subject to successful compliance reporting.
- The environmental metric developed incorporates a state-and-transition model of the relevant ecological community. This framework provides a robust ecological basis for determining both the current condition of individual assets and their likely future condition as a consequence of targeted management investments (Zammit *et al.,* in press).
- The programme incorporates independent ecological benchmarking and on-going monitoring of all investment sites to provide robust performance monitoring of the long-term ecological benefits of investments.
- The programme incorporates regular social profiling of all successful, and some unsuccessful, land managers to determine the long term impacts of the programme on individual and community values, attitudes and behaviours towards conservation management on private land.

A number of important lessons which emerged from the Fund have been incorporated into the Environmental Stewardship Program. Key lessons include:

- The design and implementation of PES mechanisms is a continuous learning process and adaptive management is essential to ensure programmes can be amended to reflect changing environmental or market conditions.
- To ensure effective on-ground delivery and credibility in the market place, on-ground delivery organisations need to have a sound track record in environmental management and an ongoing presence in the region where the PES are being run. Any on-ground delivery agent must also have the ability to maintain professional capacity and quality assurance throughout the programme delivery phase.
- To ensure effective and efficient delivery, processes need to be in place to deal with over/under subscription in PES mechanisms.
- To ensure efficient evaluation of the environmental values to be purchased, metrics need to be 'fit for purpose' and should not be over-engineered to incorporate ecological and other considerations, such as complex weighting functions, that have negligible additional discriminatory power. Sensitivity analysis is a critical component of determining 'fit-for-purpose'. Metrics also need to align with practical field assessments. In the case of Environmental Stewardship, because the target environmental communities are already protected under legislation, the metric developed explicitly focuses on the current condition of the vegetation and the likely change in condition under the proposed management arrangements.
- To reduce potential compliance risks, contracts for funding need to be longer-term to allow payments to better align with actual costs faced by landholders. Environmental Stewardship contracts run for up to 15 years.
- To ensure a robust and appropriately priced bid, providing information for participants to assist them frame and price their proposals may be necessary. This includes workshops to assist participants to understand how the PES mechanism works and information on potential tax implications of the commercial arrangements employed.

Early results for the Environmental Stewardship Programme

The first environmental asset targeted by the Stewardship programme is the critically endangered Box Gum Grassy Woodland ecological community. The PES approach is a inverse auction, similar to the Fund, but the design and implementation has incorporated the lessons from the Fund. On-ground delivery is being undertaken under contract by three non-government regional environmental organisations with substantial local environmental knowledge and established professional relationships with landholders.

To date five stewardship rounds have been undertaken. More than 500 land managers have expressed an interest in the programme and about 160 have already been successful in securing long term contract to manage over 16 000 hectares of critically endangered box gum woodlands on their properties. In addition, the competitive nature of the programme is eliciting proposals with high levels of variance in cost effectiveness enabling an efficient set of contracts to be established within the programme budget constraint.

7.5 Conclusions

The Forest Conservation Fund has been a significant application of a competitive, market driven PES mechanism for biodiversity conservation in Australia. The results achieved through the Fund have made a measurable contribution to the protection of native forest communities in Tasmania and provided a strong basis for designing and implementing future PES schemes in Australia.

A key policy lesson from the Fund is that landholders will respond differently to alternative design elements of PES schemes depending on their specific attributes. Therefore it may be worthwhile to develop and run a portfolio of different mechanisms to attract a wider mix of participants in a competitive environment.

It should also be noted that market-based approaches to achieving public good conservation outcomes are one policy tool available to policy makers. PES schemes should not be seen as a panacea or substitute but as part of the group of financial incentive tools that are increasingly available to governments to complement more traditional regulatory and suasion approaches to achieve conservation outcomes.

The inverse auction, direct offers and differentiated take-it-or-leave-it approaches all proved to be effective and efficient in securing forest protection and management in Tasmania. However, it needs to be emphasised that the efficiency of the direct offers and differentiated take-it-or-leave-it approaches used were critically dependent on price information obtained through the earlier inverse auction approaches. The application of robust statistical models provided confidence that individual take-it-or-leave-it offers were consistent with previous prices for environmental assets of comparable value.

The relative success of the Fund and Environmental Stewardship has been largely attributable to a robust design process and professionalism in on-ground implementation. However, the design and implementation of both programmes was a continuous learning process and constant monitoring and evaluation has been fundamental to improvements throughout the Fund. The lessons learned from the monitoring and evaluations of the Fund are now being applied in other Australian government PES schemes.

As environmental science and policy becomes more sophisticated and institutional arrangements change, the scope for PES schemes is widening to enable efficient payments for 'bundles of ecosystem services' that will enhance the extent and condition of multiple environmental assets (*e.g.* biodiversity, carbon, water, soil). These opportunities are currently being explored in more depth in Australia.

Notes

1. See http://www.daff.gov.au/rfa.
2. Relevant documentation on the Fund is at: http://www.environment.gov.au/land/forestpolicy/fcf/.
3. In effect, the value for money metric changed from AUD/CVI to (AUD from the proposal + future administrative AUD/CVI).
4. Because the opportunity costs of meeting the objectives of the FCF were not well understood, the program administrators did not establish a formalised acceptable maximum price (AUD/CVI). However, by establishing the cut-off for each round at the points used, this enabled funds to be withheld to purchase more cost effective bids in subsequent rounds.
5. Early analysis of the performance of the Revolving Fund indicates that it has the potential to be more cost effective than the auction approach. However, it cannot achieve large gains in conservation quickly as it is constrained by supply and demand in the existing property market. See http://www.environment.gov.au/biodiversity/incentives/revolving-funds.html.
6. It should be noted that the 12 year covenant option was removed after round 1a.
7. For this analysis, the funding budget was capped at the budget available for the actual tenders. Proposals were selected from the pool of actual proposals that only included forest types targeted by the Fund.

8. Only data from actual successful proposals was used to eliminate any very inefficient outliers from the full set of proposals. A hypothetical budget of USD 20 million was used because analysing all successful bids using the full Fund budget would result in the same aggregate outcomes (*i.e.* all of the same proposals being selected, albeit in a different order).

9. See http://www.nrm.gov.au/stewardship/index.html for details.

References

AMAP (2006), "Assessment Methodology Advisory Panel Options Paper: Methods for Assessing the Significance, Services and Security offered in Proposals to the Forest Conservation Fund and Recommendations for Tender Design", Prepared for the Forest Conservation Fund Steering Committee.

Commonwealth of Australia (2006), *Strategic Plan for the Forest Conservation Fund*, www.environment.gov.au/land/publications/forestpolicy/pubs/strategic-plan-fcf.pdf.

Commonwealth of Australia (2007), *Information kit for landholders interested in the Forest Conservation Fund*, www.environment.gov.au/land/publications/forestpolicy/fcflandholder-kit.html.

Dargavel, J. (1995), *Fashioning Australia's Forests*, Oxford University Press, Melbourne.

Eigenraam, M., P. Barker, M. Brown, R. Knight and S. Whitten (2007), "Forest Conservation Fund: Conservation Value Index Technical Report", Report to the Department of Environment, Water, Heritage and the Arts.

Marsden Jacob Associates (2010), "The Tasmanian Forest Conservation Fund and Associated Programs: Purpose, Performance and Lessons", Report to the Department of Environment, Water, Heritage and the Arts.

Productivity Commission (2004), "Impacts of Native Vegetation and Biodiversity Regulations", Report No. 29, Melbourne.

Zammit, C., S. Attwood and E. Burns (in press), "Using markets for woodland conservation on private land: lessons from the policy-research interface", in D. Lindenmayer, A. Bennett and R. Hobbs (eds.), *Temperate Woodland Conservation and Management*, CSIRO Publishing, Melbourne.

Paying for Biodiversity: Enhancing the Cost-Effectiveness of Payments for Ecosystem Services

Chapter 8

Indonesia: A pilot PES auction in the Sumberjaya watershed

Beria Leimona[1] and Brooke Kelsey Jack[2]

This chapter discusses a pilot inverse auction PES programme applied in the Sumberjaya Watershed in Indonesia to reduce sedimentation from coffee plantations. The process of design and implementation is discussed, highlighting issues that arise in a developing country context. The chapter also discusses how the pilot auction can be used as a price revelation mechanism, enabling payments to better reflect the costs of ecosystem services provision for any future scaled-up PES programme.

1. The World Agroforestry Centre – ICRAF SEA & Wageningen University and Research, The Netherlands.
2. Center for International Development, Kennedy School of Government, Harvard University, United States.

The authors thank Dr. Vic Adamowicz and Dr. Paul Ferraro for their advice and inputs to this study. This research is supported by the Economy and Environment Program of Southeast Asia (EEPSEA) and International Agriculture and Development Fund (IFAD). Elements of this case study have been previously published in Jack, Leimona and Ferraro (2008), Jack (2009), and Leimona, Jack, Lusiana and Pasha (2009).

8.1 Introduction

While inverse auctions for PES have been applied in a number of developed countries, they have to date not been widely adopted in developing countries. This chapter examines one of the few applications of inverse auctions in a rural setting of a developing country, namely in Lampung, Indonesia. A pilot PES scheme was implemented in 2006-2008 to induce farmers to reduce sedimentation in two sites in the Sumberjaya Watershed: Way Ringkih (Site 1) and Way Lirikan (Site 2). Site 1 consists of two villages Talang Kuningan and Talang Harapan, and Site 2 consists of Wanasari I and Talang Anyar. The aim of this pilot was to assess the feasibility of using auctions in a developing country context and to obtain an understanding of the drivers of farmers' willingness to accept (WTA) compensation for a conservation contract. The farmers are environmental service suppliers as they play a role in maintaining the environmental benefits from the watershed. Their decisions on land use practices influence the provision of environmental services (ES) from this landscape, including water quality, biodiversity and scenic beauty. Information on the supply curves can be valuable for designing conservation-payment programmes; estimating these costs accurately can inform conservation planners of the financial, ecological and socioeconomic implications of future scaled-up PES programmes.

As part of a PES project on the island of Sumatra led by the RUPES Phase II (Rewards for, Use of and Pro-poor Investment of Environmental Service scheme) of the World Agroforestry Centre (ICRAF), this pilot auction was implemented to elicit private information on landholders' WTA payments in return for soil conservation investments on private coffee farms. The Sumberjaya watershed is dominated by coffee crops in erosion-prone uplands. Erosion transports sediment loads to sensitive aquatic ecosystems and has serious negative effects on the resident flora and fauna. Moreover, a gradual reduction in soil organic carbon due to erosion can, depending on its deposition site, lead to a reduction in ecosystem carbon storage (van Noordwijk *et al.,* 2007). Finally, soil erosion in Sumberjaya contributes to the rapid siltation of a downstream hydropower reservoir (the PLTA Way Besai reservoir, located approximately 30km downstream of the reservoir) that provides local irrigation services and electricity for three provinces in Sumatra (Sihite, 2001; Ananda and Herath, 2003). Erosion control is an impure public good that generates both private benefits and positive externalities. As a result, farmers tend to under-invest in soil conservation.

In Sumberjaya, two approaches of rewards for environmental services schemes are introduced. First, RUPES is scaling up the forms of land tenure that are conditional on farmers maintaining environmental services, or 'tenure as reward'. The agreement for conditional land tenure is a conflict resolution tool between local people, mostly migrants from Java, and various layers of government. Second, a financial reward scheme by the hydropower company provides some funds upfront and then pays additional specified amounts based on the effected sedimentation reductions. Facilitated by RUPES, the community members learned to monitor and control the local sources of sediment in their streams and take action by establishing the River Care group. The River Care activities are a collective action to reduce sedimentation that includes the repair of the river bank, compacting of dirt paths, dredging of river mud, and building small-dams to retain sediment. One of the primary achievements of the River Care initiative was developing an easy-to-use method to link environmental service performance directly to the size of the payments. Environmental service providers can thus design effective plans for improving their performance. In doing so, they can provide greater value to external customers and earn more potential income in the process (RUPES, 2006).

8.2 Designing the PES inverse auction

Several preparatory steps were taken before the procurement auction was conducted (Figure 8.1). First, the sample population and potential auction participants were identified at the sub-watershed level. Second, the conservation contract that would be offered in the auction was designed. In designing the contract, some basic information was needed, such as: What problems would be solved by the conservation project? Do the local farmers have any knowledge in solving the watershed problems? What are these appropriate conservation techniques? What are the farmers' preferences for terms of payment? When does the contract begin? Third, some elements of the auctions were tested and selected through two types of experiments: laboratory auction experiment with students and field framed experiments with farmers.[1] The final step was to conduct a natural field experiment and monitor the contract accomplishment of farmers who obtained a contract for one year.

This study resulted in a set of auction rules to determine how the limited budget of the watershed rehabilitation fund, financed by the parastatal hydropower company, would be allocated. The watershed rehabilitation fund in Indonesia is mostly obtained from the corporations' conservation funds. The legal basis of this scheme is the Letter of Ministry of Parastatal Company Affairs over Corporate Social Responsibility Partnership Programs. It was cited that 1% of net-benefit of state-owned companies should be allocated for developing environmental programmes with the communities. This scheme could be seen as potential mechanisms for rewarding transfers through a governmental public investment scheme.

Figure 8.1. **Flow of the research steps in Indonesian pilot auction**

Participant engagement

- Identification of the potential auction participants as ecosystem service providers through a rapid socioeconomic survey.
- Assessment of watershed problems and local management options through 'transect walk', and focus group discussions.

Proposal development

- Enhancing conservation knowledge and understanding through focus group discussions and village training sessions.
- Designing conservation contracts through focus group discussions.
- Designing auction mechanism using results from conventional laboratory experiments with students and framed field auctions with farmers.

Selection process

- Implementing natural field experiment auction.
- Analysing auction data and bid behaviour to select contracts to be enrolled.

Contract implementation

- Monitoring contracts and achievements every 3 months.
- Conducting interviews with participants.
- Transferring payments.

Source: Adapted from Leimona *et al.*, 2009.

Several on-farm techniques effectively reduce soil erosion from smallholder coffee farms in the watershed (Agus *et al.*, 2002). Four focus group discussions involving 76 farmers from three villages led to the selection of three scientifically appropriate techniques: soil infiltration pits, vegetation strips and ridging between coffee trees. Farmers preferred these techniques for their suitability, familiarity and simplicity (Leimona *et al.*, 2008). All three are scalable and verifiable, and thus appropriate for contracts that make payments conditional upon performance.

Moreover, the contracted techniques reduce erosion without decreasing coffee production and incur few fixed costs, requiring primarily labour investments using tools already owned by the farmers. Components of landholders' WTA were anticipated to include both observable characteristics, such as plot slope, and unobservable characteristics, such as the opportunity cost of labour and individual discount rates. Bids in an incentive-compatible auction capture all of these factors, and thus reveal the distribution of WTA within the population.

Auction design and implementation

The socio-economic characteristics of the farmers (*i.e.* the auction participants) are: low education level (below seven years of education), low asset endowment, small plot size (mostly less than 0.5 hectares), where familiarity with market-based competitiveness is not particularly common. Several of the auction design elements were selected to respond to these characteristics and general rural situations in developing countries, where most of the participants had strong social binding among their community members, and where village leaders and elders have significant roles and dominance in decision making (Ferraro, 2004). Auction elements were chosen for their simplicity, equitable payments and transparency to ensure each participant had the freedom to reveal their own bids without any external interference. A sealed-bid auction was conducted to maintain anonymity. The second price auction was selected since it was relatively easy to explain and be understood by the participants, hence making the bidding process more transparent.

An effort-based payment mechanism was chosen because the time frame of this project was too short for accurate output-based (*i.e.* level of sedimentation reduced) performance payments. Inaccurate measurement of environmental service outcome would bias the performance achieved by the farmers and at the end, could cause any disappointment both from providers and buyers. Table 8.1 summarises the design characteristics of the auction.

To provide an incentive for truthful cost revelation, a uniform-price rule was used, where the final contract price equals the lowest rejected offer price. Under this uniform-price rule, bidders who bid above their true values cannot benefit from overbidding. This is because the price is set by the lowest rejected bid, and bidders risk losing the contract at a price they would have been willing to accept. Bidders who bid below their true value increase the likelihood of winning a contract at a price below their minimum acceptable price. Thus, all bidders' best (weakly dominant) strategy is to bid their true WTA. They can do no better, and sometimes worse, by misrepresenting their WTA. In contrast, discriminative-price procurement auctions, where winning bidders receive a contract price equal to their own bid (*e.g.* Stoneham *et al.,* 2003), or under a uniform price rule where the price is set by the last accepted offer, bidders have strategic incentives to inflate their bids to levels above their true WTA. Furthermore, Alix-Garcia *et al.* (2003) show that uniform pricing may be more equitable, while discriminatory pricing is more cost-effective (see Chapter 1).

Table 8.1. **Indonesian pilot auction design characteristics**

Characteristic	Implementation
Auction type	One-sided, sealed bid procurement auction
Bidding units	Willingness to accept (WTA)
Budget limit	Predetermined, concealed
Number of rounds	7 provisional, 1 binding
Announcement of provisional winners	By ID number
Bid timing	Simultaneous
Pricing rule	Uniform, lowest rejected price
Tie-breaking rule	Random in determining tied winners
Bidder number	Known, fixed
Activities contracted	Determined in advance

Source: Adapted from Leimona *et al.*, 2009 and Jack *et al.*, 2008.

In gametheory, a reserve price is the maximum acceptable bid.[2] For this auction, a reserve price was preset, but was not announced since the announcement of reserve prices can influence the bidding strategy (Latacz-Lohmann and Schilizzi, 2005). However, the bidders can also implicitly interpret information in their winning bids as reserve prices in multiple round auctions. To avoid bidder learning between preparatory bidding rounds, only the winning ID numbers were announced, and the total conservation budget was not revealed.

The conservation auction was carried out on consecutive days in two nearby villages in a single sub-watershed. The villages were selected based on hydrological studies showing their contribution to sediment loads. A random sample of participants from the sub-district population would have provided results more in keeping with the purposes of this study, but the interests and preferences of ICRAF to integrate its biophysical and socioeconomic research precluded this approach.

The primary occupation in the two villages is coffee farming, most of which takes place on small, individually-owned plots that are not subject to any land-use regulations. The auction was limited to owners of private coffee plots, and excluded plots on state-forest lands which are subject to other regulations. One village comprised 55 households, 53 of which owned private agricultural land. Of these, five rented or sharecropped their land, leaving 48 eligible households, all of which participated in the auction. In the other village, 55 of the 87 households owned private agricultural land. Of these, 20 rented or sharecropped their land. Thus 35 households were eligible, and 34 participated in the auction. To ensure that participants understood the contract requirements, all participating farmers attended field training. The theory and practice of erosion control management techniques were presented, and site visits were made to adjacent villages where erosion control management was already in place.

Farmers, each designated with an identification number, submitted sealed bids representing their per-hectare price for accepting a conservation contract.[3] Farmers were informed that payments would be made in three instalments, with the second two conditional upon verification of compliance. The multi-instalment payment plan provided incentives for compliance for the duration of the contract, which mitigated valuation problems associated with moral hazard (*i.e.* lowering bids because of the expectation of lax enforcement). In addition, the farmers expressed a preference for periodic payments during focus group discussions, likely due to a lack of access to credit markets. As the primary purpose of the auction was to accurately estimate supply curves (rather than to maximise the conservation benefits per dollar spent), plots were not ranked by their erosion mitigation potential. Farmers were aware that enrolment decisions were based solely on their bid price per hectare. Contracts were treated as discrete (*i.e.* either all or none of plot was contracted), though contracting could also have treated hectares as the discrete unit.

In each of the two villages, the auction lasted 2-3 hours, during which the participants heard the contract described, received instructions about the auction, and submitted their bids. Following Cummings *et al.* (2004), the auction was designed with several provisional rounds preceding the final allocation round. After each provisional round, the bidder identification numbers of provisional winners were announced. No price information was provided between rounds and participants were not allowed to converse. Bids were revised and re-submitted for each round, a process designed to increase familiarity with the mechanism (Cummings *et al.,* 2004). Participants were informed of the number of provisional rounds in advance to ensure that final round bids were based solely on WTA and not subjective expectations about the number of rounds. Jack (2009) noted that the multiple familiarisation rounds in Sumberjaya auction resulted in reduced bid inflation, thus allowing a larger land area to be enrolled – or in other words, increases the efficiency of the auction.

The contractual arrangements between the two sites were different. At Site 1, two farmer groups (one from each *talang*) signed the contracts. The members arranged working in rotation, shifting from one plot to another until all the contracted activities were finalised. At Site 2, farmers signed individual contracts with ICRAF. In other words, there were two group contracts at Site 1, and 15 individual contracts at Site 2.

8.3 Auction outcomes and environmental impacts

Of the 82 auction participants bidding on 70 ha, 34 participants received contracts for soil conservation activities on a total of 25 ha at an average price of USD 171.70 (1 USD = 9000 IDR). The total budget of around USD 4 450 was combined with the uniform pricing rule to determine the contract price of USD 177.78/ha in the first village and USD 166.67/ha in the second village. Just over one additional hectare of conservation investment would have been purchased if participants were paid their own bid (*i.e.* discriminative-price auction). However, as explained above, bid inflation under a discriminative-price rule would reduce these gains. In the following discussion, we did not consider a single high outlier bid.

Table 8.2. Indonesian pilot auction summary statistics (USD per hectare)

Number of participants	82
Number of contracts awarded	34
Number of hectares bid	70
Number of hectares contracted	25
Contract price per hectare	171.70
Mean bid per hectare	263.14
Median bid per hectare	181.67
Minimum bid per hectare	66.67
Maximum bid per hectare	2 777.78
Standard deviation	344.91

Source: Jack *et al.*, 2008.

Figure 8.2 presents the aggregate supply curve from the two villages, *i.e.* describing the number of hectares enrolled in the programme for any given price. It follows an exponential distribution with increasing marginal costs. Note that this supply curve represents short-run costs as estimated by the participants, which may change as participants learn more about the contract or the contractor. Measuring a supply curve in terms of erosion abated would be preferred over the proxy measure of hectares under soil erosion mitigation activities. Most conservation payment initiatives, including this study, measure performance by land-use activities rather than actual services supplied, because of monitoring difficulties and the risk burden for landholders (Wunder, 2007).

Figure 8.2. **Supply curve resulting from Indonesian pilot auction**

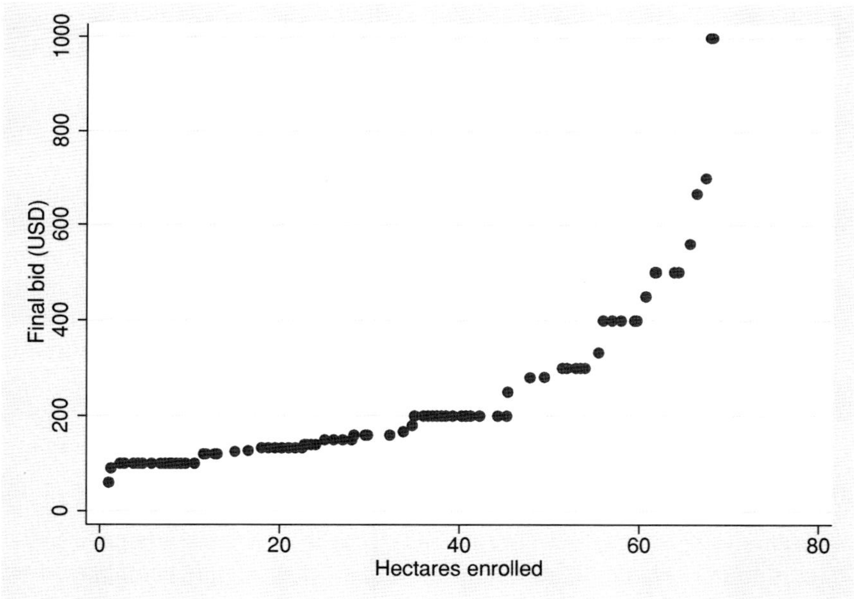

Source: Jack *et al.*, 2008.

Efficiency Gains from the Auction

To assess the efficiency of the auction, alternative methods were used to estimate the costs of the contracts prior to the auction. Labour costs were expected to comprise the primary investments needed for the contract. Labour cost information was thus elicited using two approaches. First, during focus groups, farmers were asked to estimate the labour requirements of the contract. Estimates were based on wages, number of hired workers and number of work days. The average costs approximated by the farmers were USD 300 per hectare, including forgone wages from the farmer's own labour investment. Second, cost information was collected as part of a household survey, asking about time investments for past implementation of soil conservation activities. The estimates based on retrospective calculations were slightly lower, around USD 225.

The cost estimates based on labour investments are 30 to 75% higher than the auction price of USD 171.70 per hectare, and 24 to 65% higher than the median bid. Based on estimated labour costs, 14.8 to 19.8 hectares of contracts could have been enrolled under the available budget, as opposed to the 25 hectares actually purchased under the auction (26% to 69% more). On the other hand, the mean bid price was between the two estimates based on labour costs, suggesting that these methods may have been fairly accurate in estimating mean values. This outcome does not indicate that the labour cost estimates were inaccurate, simply that they provided incomplete measures of farmers' WTA.

Contract monitoring

The research team conducted two qualitative (third and ninth month of contract signing) and quantitative (sixth and twelfth month of the contract signing) monitoring activities in the field. The qualitative monitoring obtained information on the contract implementation using open-ended questions. The enumerators checked the general quality of the conservation structure and asked farmers whether or not they had any difficulties in implementing their contacts. During quantitative monitoring, enumerators measured the size of sediment pits and observed the quality of the ridging and grass strips. They also surveyed social interactions among farmers and other conservation structures that were not required by the contract, such as water drainage and terracing. This monitoring involved two external evaluators from the District Forestry Service who independently gave scores to the farmers' accomplishments. The head of the village accompanied the team as a witness to fair evaluation. Farmers who were not able to accomplish at least 50% of the contracted activities had to give up and could not continue their contracts. At the final monitoring, the implementing agency paid the remaining fund to farmers who had accomplished at least 80% of the contracted activities.

The mid-term monitoring revealed that most farmers successfully completed their obligations. Figure 8.3 shows the average compliance for Site 1 and 2 at the six month quantitative assessment and at the end of the contract. Only one contract was terminated early; a farmer from Site 2 only achieved 4% of the required activities after six months. The exit interview revealed that the main reason for such performance was the higher opportunity cost for getting other side jobs than the contract value.[4]

Figure 8.3. **Average village compliance in Indonesian pilot within each site, measured during the middle and at the end of the contract term**

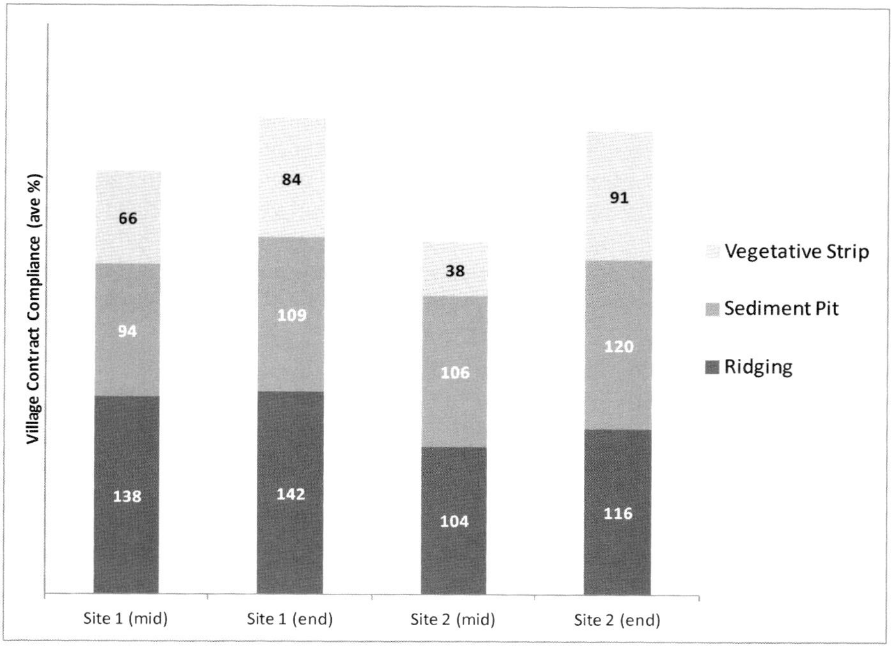

Source: Based on Leimona *et al.*, 2009.

After one-year of contract implementation, again most of the farmers showed good progress in implementing their contracts. Farmers constructed ridgings and sediment pits over and above the demands of the contract, but they lagged behind in planting the vegetative strips. Farmers also practiced other conservation techniques such as the building of terracing and drainage that could optimally support the contracted conservation efforts. All farmers constructed terracing, which could be done simultaneously with ridging and half built drainage systems.

The successful completion of planting vegetative strips was found to be influenced by other farm priorities. For example, in Talang Kuningan, Site 1, planting was successful, partly because they used it as extra fodder for their livestock (goats). However, in Talang Harapan, Site 1, the absence of livestock removed this extra incentive and less effort was put into planting vegetative strips. This highlights how conservation measures are especially successful when they are mutually advantageous for the landholders.

In summary, 19 out of 34 farmers successfully accomplished the contract requirements (*i.e.* 55% across the two sites). Fourteen farmers did not pass the final evaluation and one farmer failed during the mid-term evaluation. Most of them failed in planting the grass strips although many of them constructed both ridging and sediment pits, even exceeding the contractual agreement. We decided that for the final decision, the percentage of accomplishment would not be calculated cumulatively. We did not add up all the percentages but evaluated these individually. Thus, farmers who failed one of the contracted components were not eligible for the final payment. Although the rate of accomplishment could be categorised as low, we could not conclude that the overall conservation effort was not successful. Table 8.3 shows that the rate of accomplishment was greater than 80% for all contracted techniques: ridging (128%), sediment pits (114%), and grass strip (88%).

Table 8.3. **Rate of contract accomplishment in Indonesian pilot**

	Total number of farmers	Number of failed farmers	Rate of success (%)
Site 1	**19**	**10**	**47**
Talang Kuningan	9	0	100
Talang Harapan	10	10	0
Site 2	**15**	**6**	**67**
Wanasari I	10	4	70
Talang Anyar	5	2	60

Source: Leimona *et al.*, 2009.

Each *talang* (sub-villages) across the two sites had different rates of success in accomplishing their contracts. At Site 1, all farmers (100%) in Talang Kuningan fulfilled their contractual agreement, while in Talang Harapan, no farmer received the final payment. The rate of success at Site 2 was higher (67%) and well-distributed at each *talang* compared to Site 1, with a 47% rate of success.

The different contractual arrangements and institutions are likely to have influenced the rate of success of each *talang*.

An exit interview was conducted to examine the underlying motivations for contract performance. Most of the Talang Harapan farmers, where group contracts were issued, cited the lack of leadership and poor coordination as the major reasons why their group was not motivated in performing well. The field assistant observed that the group did not choose the leader voluntarily, and the group leader was not an active community member. Farmers also cited time-constraints as a factor, due to other activities, such as harvesting coffee, working in the rice field and other gardens, engaging as daily labourers, and renting motor bikes. Unsuitable weather was another factor. In reality, many other farmers could easily find grass and accomplish fully the conservation activities with the current weather. However, most of them felt that they could not accomplish the contract at the sixth month as this coincided with the coffee harvesting period. Some of the farmers also assumed that receiving a low score during the mid-term evaluation could influence the final result, hence lowering their motivation to complete the contract.

The farmers suggested some improvements to increase the conservation program's rate of success. At least six farmers proposed having individual contracts rather than group contracts because weak coordination among members could make the whole group fail. Some contract components should be more flexible, they said. Most of them agreed that there should be sanction and that the current sanction was suitable. None of the farmers had problems with the design of the auction and the contractual agreement. Subsequent analysis showed that there was no significant difference in conservation awareness level, understanding on the auction design (rules, complexity), information quality and level of satisfaction between farmers who complied fully with the contract and those who did not.

Environmental impact of contract implementation to sedimentation reduction

To evaluate the impacts of the PES on water sedimentation, field researchers took water samples in the two watersheds (Way Ringkih and Way Lirikan) three times: June, November, and December of 2007, at three observation points located at the final outlet of the Way Ringkih and Way Lirikan River before entering the Way Besai and at the end of Talang Kuningan stream before flowing to Way Ringkih. Sedimentation data at the first two points for the year 2005 were available for comparison.

The effect of a one-year contractual agreement to reduce river sedimentation was uncertain. In Way Ringkih, the sedimentation rate at the beginning of December 2007 was higher (1 283 milligram/litre) compared to the rates in 2005 (1 027 mg/l) to mid 2007 (528 mg/l). In Way Lirikan, the sedimentation rate in December 2007 (296 mg/l) was consistently lower than the average rate in 2005 to mid 2007 (603 mg/l). In Way Lirikan, the decrease of rate of erosion was lower than in Way Ringkih because the River Care programme activities were already being carried out in the area during the auction contract period.

The conservation activities of the auction pilot sites, however, were not the main factors that decreased the sedimentation rates. Rather, the scale of conserved land under the contract was too small, covering only 25 hectares, and the one year contract period was too short. The time lag for the real effect of erosion reduction is about 10 to 50 years for any intermediate alteration of the landscape at watershed scale.[5] Living and dead plant biomass, vegetative cover, soil structure and amount of rainfall are among the factors that can influence erosion (Verbist, 2008; Pimentel *et al.,* 1995).

8.4 Conclusions

Based on the outcomes from the laboratory and field experiments as well as theoretical considerations, the design of this pilot auction was a sealed bid auction with budget constraints, random tie-rule, uniform pricing rule, minimised collusion, announced ID numbers of provision winners and announced number of rounds. The auction followed a fairly standard format, with a single buyer and multiple sellers submitting sealed bids representing their WTA the soil conservation contract for their plot. Bids were assessed according to a per-hectare price and the cut-off price was determined by a pre-set budget constraint.

The auction for the PES programme in Indonesia was designed using a uniform price rule for equity reasons. The literature on auction design finds that uniform pricing is more likely to reveal farmers' true opportunity cost because bidders only determine the chance of winning. However, uniform pricing is relatively less cost-effective compared to the discriminative price rule.

The auction was a multiple round consisting of eight rounds with the last binding round. The benefit of multiple rounds was that farmers learned from the rounds of the auction. However, the announced last round may introduce forms of strategic behaviour. Concealing the number of rounds will give participants higher uncertainty because they have their own subjective probability distribution about the chance of the last round. By announcing the last round, the benefits from farmers' learning on the previous round and the advantages of a one-shot auction for the last round were combined.

The rate of accomplishment at the final monitoring was moderate. The reasons for this were various, ranging from lack of leadership and coordination among farmer group members, difficulty in finding grass seedlings to accomplish the contract, and coincidence with coffee harvesting time. In this specific case, private contract tends to be more successful compared to collective contract when leadership is lacking or 'champion' among the community members does not exist. Institutional aspects and contract flexibility might influence the accomplishment of conservation efforts. Analysis showed that there were no significant differences in level of understanding, complexity, and competitiveness and conservation awareness between compliant and non-compliant farmers.

A limitation of this study is that all units of the pilot site were treated as homogeneous, with respect to their contribution to erosion and downstream sedimentation. These sites' contribution to environmental services is also heterogeneous, related to hydrological and geophysical factors that are unlikely to be correlated with cost. The emphasis of this pilot auction was to assess the feasibility of the auction approach in a developing country context and to obtain an understanding of farmers WTA and the drivers thereof. A scoring rule giving higher values to plots that contribute more to downstream problems is preferable. For instance, plots located on steeper slopes and closer to rivers and streams could be assigned higher values so as to enhance the cost-effectiveness of a larger scale auction. The simplifications in this pilot auction were deemed appropriate for the research and valuation intentions of the study. For a larger scale allocation auction, modifications such as using supply curve information resulting from this procurement auction would be more appropriate. Such valuation information provides a reasonable platform for designing a scaled-up fixed payment scheme, including differential rates and eligibility rules necessary for targeting participants.

The design of an experimental auction should fit the purpose of overall objectives of a conservation programme. In this case, the challenge was to design and administer a fair auction for farmers with low formal education, prone to social conflicts, and influenced by power structures within their community.

Notes

1. This taxonomy of field experiments proposed by Harrison and List (2004) differentiated between field experiments from conventional lab experiments: A *conventional lab experiment* is "one that employs a standard subject pool of students, an abstract framing, and an imposed set of rules".

A *framed field experiment* is an experiment that "employs a nonstandard subject pool with field context in either the commodity, task, or information set that the subjects can use".

A *natural field experiment* is "the same as a framed field experiment but where the environment is one where the subjects naturally undertake these tasks and where the subjects do not know that they are in an experiment".

2. Shor, Mikhael, "Reserve Price" Dictionary of Game Theory Terms, Game Theory .net, *<http://www.gametheory.net/dictionary/ url_of_entry.html>* Web accessed: June 06, 2008.

3. Farmers had to reveal an average willingness to accept per hectare, rather than a different price for each hectare of their property, because we believed farmers would have found varying prices per hectare confusing and because uniform-price auctions in which bidders bid multiple units are not necessarily incentive-compatible (Ausubel, 1996).

4. Because of his lower economic condition compared to others and his small landholding of only 0.5 hectare, he had to spend most of his time working as a farm labourer, hence giving him little time to manage his own coffee garden. However, he affirmed that the auction was fair and that the conservation program was important in motivating farmers to conserve their lands.

5. Dillaha, T. (2007). Monitoring Changes in Hydrologic Response due to Land Management Changes at the Watershed Scale: Time Lag and Other Issues. Presented at the Global Event on Payment/Reward for Environmental Services, Mataram, Indonesia, 22-27 January 2007.

References

Agus, F., A. Gintings and M. Van Noordwijk (2002), *Choices of agroforestry and soil conservation techniques for coffee farming in Sumberjaya, Lampung Barat, Indonesia.* World Agroforestry Centre, Bogor, Indonesia.

Alix-Garcia, J., A. de Janvry and E. Sadoulet (2003), "Targeting Payments for Environmental Services: The Role of Risk", *Agricultural and Resource Economics Update* 7(4).

Ananda, J. and G. Herath (2003), "Soil erosion in developing countries: a socio-economic appraisal", *Journal of Environmental Management*, Vol. 68.

Ausubel, L. M. (1996), "An Efficient Ascending-Bid Auction for Dissimilar Objects", Working Paper No. 97–06, University of Maryland, Department of Economics.

Cummings, R G., C.A. Holt and S.K. Laury (2004), "Using laboratory experiments for policymaking: an example from the Georgia irrigation reduction auction", *Policy Analysis and Management*, Vol. 23, No. 2.

Dillaha, T. (2007), "Monitoring Changes in Hydrologic Response due to Land Management Changes at the Watershed Scale: Time Lag and Other Issues", Presented at Global Event on Payment/Reward for Environmental Services, Mataram, Indonesia, 22–27 January 2007.

Ferraro, P. (2004), "Direct Payment to Protect Endangered Ecosystems and Experimental Methods to Estimate Payment Costs", A paper for the 21st Biannual Workshop of Economy and Environment Program for Southeast Asia (EEPSEA), 16–20 May 2004.

Harrison, G. W. and J. List (2004), "Field Experiments", *Journal of Economic Literature*, Vol. 42, No. 4.

Jack, B.K. (2009), "Auctioning conservation contracts in Indonesia– Participant learning in multiple trial rounds", CID Graduate Student and Research Fellow Working Paper No. 35. Center for International Development at Harvard University, February 2009.

Jack, B.K., B. Leimona, P.J. Ferraro (2008), "A Revealed Preference Approach to Estimating Supply Curves for Ecosystem Services: Experimental field auctions and soil erosion control in Indonesia." *Conservation Biology,* Vol. 23, No. 2.

Latacz–Lohmann, U. and S. Schilizzi (2005), "Auctions for conservation contracts: a review of the theoretical and empirical literature", Report to the Scottish Executive Environment and Rural Affairs Department.

Leimona B., B.K. Jack, R. Pasha and S. Suyanto (2008), *A field experiment of direct incentive scheme for provisioning watershed services,* Environment and Economy Program for Southeast Asia (EEPSEA), Singapore.

Pimentel, D., C. Harvey, P. Resosudarmo, K. Sinclair, D. Kurz, M. McNair, S. Crist, L. Shpritz, L. Fitton, R. Saffouri and R. Blair (1995), "Environmental and economic costs of soil erosion and conservation benefits", *Science*, Vol. 267.

RUPES (Rewarding Upland Poor for Ecosystem Services) (2006), *RUPES Sumberjaya Brief No. 2,* World Agroforestry.

Sihite, J. (2001), "Evaluasi dampak erosi tanah model pendekatan ekonomi lingkungan dalam perlindungan DAS: kasus sub-DAS Besai DAS Tulang Bawang Lampung", Southeast Asia Policy Research Working Paper No. 11.

Stoneham, G., V. Chaudhri, A. Ha and L. Strappazzon (2003), "Auctions for conservation contracts: an empirical examination of Victoria's BushTender Trial", *The Australian Journal of Agricultural and Resource Economics*, Vol. 47.

van Noordwijk, M., B. Leimona, L. Emerton, T.P. Tomich, S.J. Velarde, M. Kallesoe, M. Sekher and B.M. Swallow (2007), "Criteria and Indicators for Environmental Service Compensation and Reward Mechanisms: Realistic, Voluntary, Conditional and Pro-Poor", ICRAF Working Paper No. 37, World Agroforesty Centre, Nairobi, Kenya.

Verbist, B. and R.H. Widodo (2008), *Deforestation or Climate Change: What is Changing the Flow regime of the Way Besai?,* Bogor, Indonesia: World Agroforestry Centre (ICRAF, SEA Regional Office).

Wunder, S. (2007), "The Efficiency of Payments for Environmental Services in Tropical Conservation", *Conservation Biology*, Vol. 21, No. 1.

Paying for Biodiversity: Enhancing the Cost-Effectiveness of Payments for Ecosystem Services

Chapter 9

Conclusions

This chapter highlights the key policy-relevant outcomes and lessons learned from across the book to enhance the cost-effectiveness of current and future Payments for Ecosystem Services programmes. In particular, the key criteria for effective PES are summarised and the main design elements of the three in-depth PES case studies reviewed in the book are compared.

Drawing on analysis and more than 30 PES case studies, this book has presented concepts, methods and tools to enhance the cost-effectiveness of such programmes. It aims to offer insights for good practice in PES design and implementation, including how to target the available resources so as to achieve the greatest biodiversity and ecosystem service benefits at low cost and how to best mobilise finance for PES. Looking at three case studies in particular, the book also examines experience with the design and implementation of inverse auctions in PES. Inverse auctions are innovative approaches to enhance PES cost-effectiveness and are becoming increasingly and successfully applied across developed and developing countries.

PES programmes aim to enhance the provision of ecosystem services by compensating landholders for the additional costs of providing those services. Such payments are needed to help address the externalities associated with biodiversity and ecosystem services and the fact that they often display public good characteristics. PES programmes are one policy instrument available to decision-makers for achieving positive environmental outcomes. They should not be seen as a panacea or substitute to other mechanisms but rather as part of a policy mix of incentive tools that are available to governments to complement more traditional regulatory approaches used to achieve environmental objectives. PES are based on a system where the user or beneficiary pays for the ecosystem services they would like to benefit from. The choice of the appropriate instrument will depend on distributional concerns, and the allocation of property rights that establish the "reference level" defining who should pay and who should be paid for the provision of ecosystem services. In developing countries, ecosystem service providers are generally thought to be poorer than the service users, thus creating an equity argument for positive incentive-based approaches (Pagiola *et al.,* 2005). Agri-environment payments for example are used in several developed countries, such as EU countries, Norway, Switzerland, the United States (Vojtech, 2010). In developed countries therefore, interest in PES may continue to increase as governments consider ways to re-orient existing policies so as to better promote environmental objectives. The lessons and insights from PES may be particularly relevant in the context of the EU Common Agricultural Policy reform, for example.

The explicit recognition of use of PES in ecosystem restoration programmes is also likely to be helpful to CBD parties, who agreed at the Nairobi implementation meeting (recommendation 3/6) to include an item on the CBD COP-11 agenda (probably in 2012) on "The identification of ways and means to support ecosystem restoration, including the possible development of practical guidance on ecosystem restoration and related issues".

PES are direct incentive-based instruments with potentially large gains in cost effectiveness compared to indirect payments or other regulatory approaches for biodiversity and ecosystem service conservation and sustainable use. The degree to which cost effectiveness can be achieved however depends crucially on PES programme design and implementation. While the optimal design of the programme is dependent on the specific goals, priorities and context of the programme, there are however common principles and criteria that underlie any effective PES programme. The key criteria to be considered are summarised below:

1. *Remove perverse incentives*: For a PES programme to produce clear and effective incentives any conflicting market distortions must be removed.

2. *Establish clear and enforceable property rights*: The ecosystem service provider must have clearly defined and enforceable property rights over the land providing the services.

3. *Clearly define PES goals and objectives*: Clear objectives will help guide the design of the PES programme, enhance transparency, and can minimise *ad-hoc* political influence.

4. *Develop a robust monitoring and reporting framework*: Monitoring, reporting and verification of PES is fundamental, enabling the assessment and hence improvement of programme performance over time.

5. *Establish baselines to ensure additionality of ecosystem service benefits*: A PES programme should only make payments for ecosystem services that are additional to the business-as-usual baseline. It is essential to target payments to those ecosystem services that are at risk of loss or degradation, or that payments lead to management practices that enhance the provision of ecosystem services.

6. *Identify buyers and ensure sufficient and long-term sources of financing*: Whether the buyers of services are the beneficiaries themselves, or a government or institution acting on behalf of them, the finance must be sufficiently predictable and long-term to ensure that the objective of the PES can be met.

7. *Identify sellers and target ecosystem service benefits*: Accounting for spatial variation in ecosystem service benefits via economic valuation, scoring and benefit indices, and mapping tools can substantially increase the environmental and cost-effectiveness of the programme, targeting and prioritising those sellers that offer the greatest benefits per unit cost.

8. *Consider bundling or layering multiple ecosystem services*: Bundling and layering can provide opportunities to increase the aggregate benefits

of the programme, while reducing transaction costs. Potential trade-offs in the supply of different types of ecosystem services must be identified.

9. *Address leakage*: Leakage is likely to be a problem if the provision of ecosystem service in one location increases pressures for conversion in another. If leakage risk is expected to be high, the scope of the monitoring and accounting framework may need to be expanded to enable its assessment and measures introduced to address it accordingly.

10. *Ensure permanence*: Events such as forest fires or illegal logging may undermine the ability of a landholder to provide an ecosystem service as stipulated in a PES agreement. If these risks are high, this will impede the effective functioning of a PES market. Insurance mechanisms can be introduced to address this.

11. *Reflect ecosystem providers opportunity costs via differentiated payments*: In addition to targeting payments to those ecosystem services with highest benefits and highest risk of loss, differentiated payments, equivalent to the opportunity costs of ecosystem service supply, can significantly enhance PES cost-effectiveness. Inverse auctions are one way to implement a differentiated payment mechanism – such auctions are now being increasingly and successfully applied in a number of PES programmes.

12. *Deliver performance-based payments and ensure adequate enforcement*: Ideally payments should be ex-post, conditional on the ecosystem service provision. Where this is not feasible, effort based payments are an acceptable second best, provided that changes in ecosystem management practices will bring about the desired change in service provision.

Some of the key design elements of the three in-depth case studies reviewed in this book, namely the US Conservation Reserve Programme, the Tasmanian Forest Conservation Fund, and the pilot PES implemented in the Sumberjaya district in Indonesia, are summarised in following Table.

Table 9.1. Key design elements in three PES case studies

	USDA Conservation Reserve Programme	Tasmanian Forest Conservation Fund	Sumberjaya Pilot Auction
Purpose and Scope			
Objective	Reduce environmental impacts of agriculture	Protect old growth forests on private land	Estimate opportunity costs of changing coffee farming practices to reduce erosion
Principle Ecosystem Services	Erosion control; water quality; air quality; wildlife habitat	Environmental quality; biodiversity	Water quality
Programme Scale	National	Regional	Local
Monitoring and Reporting Framework			
Monitoring	Random spot checks of 1% of contracts annually	Landholders required to report on management actions. Ongoing monitoring through Tasmanian Government	Conducted every 3 months, with farmer interviews
Targeting Benefits and Costs			
Ecosystem Benefits	Intrinsic quality of land and proposed management practices are evaluated via the Environmental Benefits Index, which scores multiple ecosystem service benefits. Landholders submit information	Multiple aspects of forest quality and proposed management practices are included in the Conservation Value Index (CVI), calculated during site visits by officials	Not targeted. Assumed to be constant across area
Risk of Loss and Additionality	Eligibility requirements used to contribute to additionality	CVI includes an assessment of current condition of the proposal areas based on benchmarked forest conditions and a Regional Threat Index to proposed forest areas	Management practices are considered additional

Table continued over page

Key design elements in three PES case studies
(cont.)

	USDA Conservation Reserve Programme	Tasmanian Forest Conservation Fund	Sumberjaya Pilot Auction
Opportunity Costs	Inverse auction (discriminatory-price)	Inverse auction (discriminatory-price auction, and modelled take-it-or-leave-it offers)	Inverse auction (uniform price)
Payment Mechanism and Contract			
Payment Source	Public funds allocated periodically in Farm Bill	Public funds allocated in Tasmanian Regional Forest Agreement	NGO grant
PES Contract and Length	Land retirement and management activities; 10 to 15 years	Legal protection and management activities; 12, 24, 48 years, and in-perpetuity	Land management activities; 1 year
Payment Mode and Amount	Differentiated payments USD 13 - 398 per ha (2009, general and continuous sign-up) Annual payments based on effort	Differentiated payments AUD 925 – 1683 per ha Ex-ante, 20% on signing contract, 80% on signing of covenant	Uniform payments USD 171.70 per ha Ex-post, dependent on completion of tasks
Assessment and Review			
Programme Revisions	Ongoing every few years via Farm Bill, particularly changes to targeting and ecosystem service priorities	Ongoing evaluation undertaken by Tasmanian Government. No new contracts are being signed	No new contracts are being signed

The issue of targeting ecosystem service payments is the main determinant in enhancing the cost-effectiveness of PES. As highlighted, the greater the spatial heterogeneity in costs and benefits of ecosystem service provision, the larger the gains that can be reaped by targeting and differentiating payments accordingly. The three elements that vary spatially are the benefits of ecosystem services, the risk of loss or degradation, and the opportunity costs associated with providing those services. Indeed, new and innovative approaches to targeting ecosystem services are being developed and applied in PES programmes, several of which have been facilitated in part by technology innovation such as GIS and satellite imagery. Though biodiversity benefits are particularly difficult to target (in comparison for example with carbon-related ecosystem services, where a clear metric, tCO_2e, is available), there are increasingly more programmes and initiatives that are available from which lessons can be learned. As discussed in the U.S. Conservation Reserve Programme case study for example, an Environmental Benefit Index is used to help target and prioritise payments to agricultural lands that can offer multiple environmental benefits. These include wildlife habitat cover benefits, as well as water and air quality benefits, amongst others. The use of the EBI, and the allocation of a maximum number of points across the different environmental factors that comprise the total EBI score, help to allocate contracts in an objective and transparent manner. Given the size and scope of this national agri-environment programme, which covers highly heterogeneous environments, one trade-off that has been noted in the design of the EBI target is that though it helps to select sites that offer a well-rounded suite of environmental benefits, it therefore discriminates against sites offering exceptional benefits in one category, but few benefits in other categories. Other complementary conservation programmes in the United States however focus on specific high-quality sites and take local and regional environmental priorities into account. These programmes therefore help to offset some of the generalities of the national EBI targeting mechanism. An alternative way to help offset these generalities would be to modify the EBI category point weighting by location (see Chapter 6).

A similar type of index, namely the Conservation Value Index, is used to help target sites with high biodiversity benefits in the Tasmanian Forest Conservation Fund, in Australia. This programme was put in place in 2005 and is of a regional scope. As such, policy design decisions of the FCF were also supplemented by GIS and ecosystem mapping. The CVI incorporates several considerations including a forestry priority score; an assessment of the current condition of proposed areas based on benchmark forest conditions; a regional threat index (these latter two which are a form of a baseline); and the likely impacts of any voluntary conservation management activities on improving conditions. In this programme, the use of the

Conservation Value Index alone is estimated to provide an 18.6% gain in conservation outcomes.

The use of econometric models, as illustrated in the Mexican PEHS, can be used to estimate the risk of ecosystem service loss. To be additional, payments must only be made to those ecosystem services that are at risk of degradation or loss, or to enhance their provision. Identifying the opportunity costs of ecosystem service provision, so as to target and differentiate payments, can be undertaken using costly-to-fake signals (as was done in the design of the Madagascar PES) or via the use of inverse auctions.

Results from applications of inverse auctions demonstrate that they can lead to large cost-effectiveness gains. In Australia for example, the inverse auction mechanism applied in the Tasmanian Forest Conservation Fund programme resulted in a 52% cost-efficiency gain (compared to a first-come-first-served approach to allocating contracts). Likewise in the United States, a local PES programme in the Conestoga watershed found that the use of inverse auctions resulted in a seven-fold increase in the reduction of phosphorus runoff per dollar spent compared to a fixed price approach (Selman *et al.*, 2008).

Though inverse auctions are gaining attention in the policy agenda and their application is becoming increasingly widespread, concerns have been raised in the context of some other programmes (*e.g.* the Scottish Challenge Fund – see Chapter 2), that landholders have perceived differentiated payments as unfair. In the case of fixed budgets for PES programmes, a situation which is often prevalent, differentiating payments so as to reflect opportunity costs implies a trade-off between larger payments for fewer people and smaller payments for more people. From a distributional point of view therefore, it is not clear which is more desirable (Ferraro, 2008). Moreover, inverse auctions have been used in several other contexts such as oil and gas in Canada and Russia, and timber and forest products in Bhutan, Costa Rica, India and Thailand (Ferraro, 2008).

In cases where there may still be social and political impediments to implementing inverse auctions, it is important to note that pilot auctions can nevertheless be used as an effective price-revelation mechanism, to help inform the design of a scaled-up uniform price PES programme. The case study of the pilot inverse auction applied the Sumberjaya watershed in Indonesia illustrates that inverse auctions can also successfully be used in developing countries to help inform the design of any future large-scale PES.

Finally, a robust monitoring and evaluation framework is fundamental to the success of a PES programme. Many long-standing and recent PES programmes that are currently in place are continuously revising and adjusting programme design and implementation so as to more cost-effectively capture the potential ecosystem service benefits. This is clearly seen in a number of programmes, including the Mexican PEHS, the Tasmanian FCF in Australia, and the US CRP. PES programmes entail a continuous learning process and a comprehensive monitoring and evaluation framework is essential to allow for improvements throughout the programme lifetime.

References

Ferraro, P. (2008), "Asymmetric information and contract design for payments for environmental services", *Ecological Economics*, Vol. 65, No. 4.

Pagiola, S., A. Arcenas and G. Platais (2005), "Can payments for environmental services help reduce poverty? An exploration of the issues and the evidence to date from Latin America", *World Development*, Vol. 33.

Selman, M., S. Greenhalgh, M. Taylor, and J. Guiling (2008), "Paying for environmental performance: potential cost savings using a reverse auction in program sign-up", World Resources Institute, Policy Note No. 5, Washington DC.

Vojtech, V. (2010), "Policy Measures Addressing Agri-environmental Issues", *OECD Food, Agriculture and Fisheries Working Papers*, No. 24, OECD Publishing. doi: 10.1787/5kmjrzg08vvb-en.

Paying for Biodiversity: Enhancing the Cost-Effectiveness of Payments for Ecosystem Services

Annex A

Case study overview

Table A.1. Case study overview

Country	Programme	Objective	Targeting Ecosystme Service payments			Location in book
			Ecosystem Service Benefits	Risk of Loss (or method to address additionality)	Opportunity Costs	
Australia	Tasmanian Forest Conservation Fund	Forest conservation	Yes. Conservation Value Index	To some extent. Risks of non-additionality included in CVI	Yes. CVI per unit cost, via auction	Section 4.1 Table 4.1 Chapter 7
Australia	Environmental Stewardship Programme	Environmental quality	Yes. Conservation Value Index	Change in management practices considered additional to business as usual	Yes. CVI per unit cost, via auction	Section 7.4
Australia	Victorian BushTender	Native vegetation conservation	Yes. Biodiversity Benefits Index	Change in management practices considered additional to business as usual	Yes. BBI per unit cost, via auction	Section 3.1 Table 4.1
Austria	OPUL	Agri-environmental quality	Not explicitly. Payments made for different management practices by area	Change in management practices considered additional to business as usual	No. Uniform payments for given management practices	Section 2.1 Table 4.1
Brazil	Ecological Value-Added Tax	Hydrological services	Includes numerous different projects	Includes numerous different projects	Includes numerous different projects	Section 4.1
Bulgaria and Romania	Danube	Biodiversity, environmental quality	Includes numerous different projects	Includes numerous different projects	Includes numerous different projects	Section 4.2 Table 4.1
Cambodia	Tmatboey	Avian species protection	To some extent. Two tiers of payments based on species viewings	Not explicitly	No. Uniform payments. Opportunity cost heterogeneity is not considered	Section 4.2 Table 4.1
Canada	Assiniboine River watershed	Wetlands and waterfowl protection	Yes. Waterfowl productivity potential estimated	Restoration considered additional to business as usual	Yes. Benefits per unit cost, via auction	Section 3.1 Table 4.1
China	Sloping Land Conversion Program	Erosion control	No. Payments per unit area	Not explicitly	No. Uniform payments. Opportunity cost heterogeneity not considered	Section 2.2
Costa Rica	Payments for Environmental Services	Forest conservation, hydrological services	Not explicitly. Eligibility criteria outline priority areas	Not explicitly	No. Uniform payment for given management practices	Section 2.1, 2.2 Section 3.1, 3.3 Section 4.1, 4.2, 4.3

Table A.1. continued over page

Table A.1. Case study overview
(cont.)

Country	Programme	Objective	Targeting Ecosystme Service payments: Ecosystem Service Benefits	Targeting Ecosystme Service payments: Risk of Loss (or method to address additionality)	Targeting Ecosystme Service payments: Opportunity Costs	Location in book
Dominican Republic	Upper Sabana Yegua	Hydrological services, biodiversity, carbon	Includes numerous different projects	Includes numerous different projects	Includes numerous different projects	Table 4.1
Ecuador	Pimampiro programme	Hydrological services	To some extent. Three tired payments for different forest type	Land use changes are considered to be additional	No. Uniform payments. Opportunity cost heterogeneity is not considered	Section 4.1 Table 4.1
Ecuador	PROFAFOR, FACE	Carbon sequestration	To some extent. Cost-environmental benefit trade-offs considered in contract selection.	Land use changes are considered to be additional	To some extent. Cost-environmental benefit trade-offs considered in contract selection.	Table 4.1
Ecuador	Socio Bosque Project	Forest conservation	Preference is given to high quality areas, poverty also targeted	Land use changes are considered to be additional	To a certain extent. Uniform payments per ha, but additional payment increases as land area increases	Section 5.1
EU	Natura 2000	Environmental quality, biodiversity	Includes numerous projects	Change in management practices considered additional to business as usual	No. Uniform payments for given management practices	Section 3.1
France	Nestle - Vittel	Water quality	To some extent. Area major consideration	Change in management practices considered additional to business as usual	To some extent, via negotiation	Section 4.2 Table 4.1
France	Danone-Evian	Water quality, environmental quality	To some extent. Area major consideration	Change in management practices considered additional to business as usual	To some extent, via negotiation	Section 4.2 Table 4.1
Germany	North Rhine-Westphalia Pilot Tender	Grassland conservation	No. Payments per unit area	Not explicitly, pilot	Yes. Area per unit cost, via auction	Section 3.3

Table A.1. continued over page

Table A.1. Case study overview
(cont.)

Country	Programme	Objective	Targeting Ecosystme Service payments			Location in book
			Ecosystem Service Benefits	Risk of Loss (or method to address additionality)	Opportunity Costs	
Greece	Amfissa	Landscape quality	No. Payments per unit area	Area protection considered additional to business as usual	No. Uniform payment for given management practices	Table 4.1
Guatemala	Sierra de las Minas	Hydrological services	High, medium and low value water supply area identified	Land use changes are considered to be additional	No. Uniform payment for given management practices	Section 4.2 Table 4.1
India	Oach-Kuhan catchment	Hydrological services	Project area targeted, but benefit heterogeneity amongst landholders not considered	Baseline assessed. Land use changes are considered to be additional	To some extent. Opportunity costs considered to set uniform payment level, heterogeneity not considered	Section 4.1 Table 4.1
Indonesia	Krakatau Steel	Hydrological services	No. Payments per unit area	Land use changes are considered to be additional	No. Uniform payments. Opportunity cost heterogeneity is not considered	Section 4.1 Table 4.1
Indonesia	Sumberjaya watershed	Erosion control	No. Principle aim of pilot is to discover service supply curve	Land use changes are considered to be additional	To some extent. Land use changes are considered to be additional	Section 2.1 Table 4.1 Chapter 8
Japan	Kanagawa Prefecture	Biodiversity and hydrological services	Includes numerous different projects	Includes numerous different projects	Includes numerous different projects	Table 4.1
Kenya	Arabuko Sokoke Forest	Forest conservation, Biodiversity	Targets areas supplying key ecosystem services	Wood plots and restoration considered additional	Various methods of rewards are used. Opportunity cost heterogeneity is not considered	Table 4.1 Section 4.2
Kenya	Sasumua	Water quality	(Planning state)	(Planning state)	(Planning state)	Table 4.1
Madagascar	Academic study	Hydrological services, biodiversity, carbon	Yes. Environmental benefits spatially mapped	Yes. Additionality gradient estimated	Yes. Opportunity cost heterogeneity considered to rule out high-cost areas	Section 3.1, 2, 3
Mexico	Payments for Environmental Hydrological Services	Forest conservation, hydrological services	To some extent. Two tired payments by forest type	Yes. Risk of deforestation modeled for spatial targeting	To some extent. Opportunity costs considered in payment level, but uniform payments set	Section 2.1 Section 3.1 Table 4.1

Table A.1. continued over page

Table A.1. Case study overview
(cont.)

Country	Programme	Objective	Targeting Ecosystme Service payments			Location in book
			Ecosystem Service Benefits	Risk of Loss (or method to address additionality)	Opportunity Costs	
Nepal	Kulekhani Watershed,	Forest conservation	Not explicitly	Land use changes are considered to be additional	No. Negotiated payments. Opportunity costs not considered	Section 2.1
Panama	ForestRE	Hydrological services	No. Payments per unit area	Land use changes are considered to be additional	No. Uniform payments. Opportunity cost heterogeneity is not considered	Section 2.2 Section 3.1 Section 4.2 Table 4.1
Scotland	Scottish Challenge Fund	Forest conservation	Yes. Environmental Benefits Index	Yes. Afforestation considered additional to business as usual	Yes. EBI per unit cost, via auction	Section 2.2
Switzerland	Ecological compensation areas	Agri-environmental quality	Not explicitly. Payments made for different management practices by area	Changes in management practices considered additional to business as usual	No. Uniform payments for given management practices	Table 4.1
Sweden	Nordic Shell Holdings	Water quality	Yes. Water filtration achieved	Yes. Performance based payments	No. Uniform payments per weight of pollutants filtered	Section 2.2 Section 4.2 Table 4.1
Sweden	Sami villages scheme	Carnivore protection	Yes. Species reproductive success achieved	Yes. Performance based payments	No. Uniform payments irrespective of village herd losses from predation	Section 2.2
Tanzania	Eastern Arc Mountains	Forest conservation, biodiversity	Targets areas supplying key ecosystem services	Land use changes are considered to be additional	Various methods of rewards are used. Opportunity cost heterogeneity is not considered	Section 4.1 Box 4.2 Table 4.1
UK	Rural Development Programme	Agri-environmental quality	Not explicitly. Payments made for different management practices by area	Changes in management practices considered additional to business as usual	No. Uniform payments for given management practices	Table 4.1

Table A.1. continued over page

Table A.1. Case study overview

(cont.)

Country	Programme	Objective	Targeting Ecosystme Service payments			Location in book
			Ecosystem Service Benefits	Risk of Loss (or method to address additionality)	Opportunity Costs	
US	Conservation Reserve Program	Agri-environmental quality, biodiversity, carbon, water quality	Yes. Environmental Benefits Index	To some extent. Changes in management practices considered additional to business as usual	Yes. Cost factor included in EBI, via auction	Table 4.1 Chapter 6
US	Wetlands Reserve Program	Hydrological services	To some extent, eligibility criteria, enrolment on case by case basis	Wetland restoration considered additional to business as usual	To some extent, enrolment on case by case basis	Section 6.1
US	Environmental Quality Incentives Program	Agri-environmental quality	To some extent, eligibility criteria, enrolment on case by case basis	Changes in management practices considered additional to business as usual	To some extent, enrolment on case by case basis	Section 2.2 Section 6.1
US	Conservation Stewardship Program	Agri-environmental quality	To some extent, eligibility criteria, enrolment on case by case basis	Changes in management practices considered additional to business as usual	To some extent, enrolment on case by case basis	Section 6.1
Wales	Tir Gofal	Agri-environmental quality	Not explicitly. Payments made for different management practices by area	Changes in management practices considered additional to business as usual	No. Uniform payments for given management practices	Table 4.1

Source: OECD, 2010.

OECD PUBLISHING, 2, rue André-Pascal, 75775 PARIS CEDEX 16
PRINTED IN FRANCE
(97 2010 11 1 P) ISBN 978-92-64-09026-2 – No. 57541 2010